Design, Assessment, Monitoring and Maintenance of Bridges and Infrastructure Networks

Relevant advances have been accomplished by the scientific community and engineering profession in the design, assessment, monitoring, maintenance and management of sustainable and resilient bridge structures and infrastructures. These advances have been presented and discussed at The Sixth International Conference on Bridge Maintenance, Safety And Management (IABMAS 2012), held in Stresa, Italy from 8 to 11 July 2012 (http://www.iabmas2012.org). IABMAS 2012 has been organised on behalf of the International Association for Bridge Maintenance And Safety (IABMAS) under the auspices of Politecnico di Milano.

This book collects the extended versions of selected papers presented at IABMAS 2012 and invited papers originally published in a special issue of *Structure and Infrastructure Engineering*. These papers provide significant contributions to the process of making more rational decisions in bridge design, assessment, monitoring and maintenance. The editors would like to thank the authors for their contributions and hope that this collection of papers will represent a valuable reference for scientific research and engineering applications in the fields of design, assessment, monitoring and maintenance of bridges and infrastructure networks.

Fabio Biondini is a Professor of Structural Engineering at Politecnico di Milano, Italy. He is Founding Member and Executive Board Member of the International Association for Life-Cycle Civil Engineering (IALCCE) and Executive Committee Member of the International Association for Bridge Maintenance And Safety (IABMAS). He is the author/co-author of over 300 scientific publications in the fields of life-cycle civil engineering, structural reliability, structural optimization and earthquake engineering. He is the recipient of several honors and awards from several professional organizations, including IABMAS, IALCCE, IABSE, IFIP. He is a Fellow of the American Society of Civil Engineers (ASCE) and a Fellow of the Structural Engineering Institute (SEI) of ASCE.

Dan M. Frangopol is the inaugural holder of the Fazlur R. Khan Endowed Chair of Structural Engineering and Architecture at Lehigh University, Bethlehem, USA. He is the Founding President of the IABMAS and IALCCE. He has authored/co-authored 3 books and over 370 articles in archival journals including 9 prize-winning papers. He is the Founding Editor of *Structure and Infrastructure Engineering* and of the book series *Structures and Infrastructures*. He is the recipient of several medals, awards and prizes, from ASCE, IABSE, IASSAR and other professional organizations, such as the OPAL Award, Newmark Medal, Ang Award, T.Y. Lin Medal, Khan Medal and Croes Medal (twice), to name a few. He holds 4 honorary doctorates and 13 honorary professorships from major universities. He is a Foreign Member of the Academia Europaea and the Royal Academy of Belgium, an Honorary Member of the Romanian Academy of Technical Sciences, and a Distinguished Member of ASCE.

Design, Assessment, Monitoring and Maintenance of Bridges and Infrastructure Networks

Edited by
Fabio Biondini and Dan M. Frangopol

Routledge
Taylor & Francis Group

LONDON AND NEW YORK

First published 2019
by Routledge
2 Park Square, Milton Park, Abingdon, Oxon, OX14 4RN

and by Routledge
52 Vanderbilt Avenue, New York, NY 10017

First issued in paperback 2020

Routledge is an imprint of the Taylor & Francis Group, an informa business

© 2019 Taylor & Francis

Publisher's Note
The publisher accepts responsibility for any inconsistencies that may have arisen during the conversion of this book from journal articles to book chapters, namely the possible inclusion of journal terminology.

Disclaimer
Every effort has been made to contact copyright holders for their permission to reprint material in this book. The publishers would be grateful to hear from any copyright holder who is not here acknowledged and will undertake to rectify any errors or omissions in future editions of this book.

British Library Cataloguing-in-Publication Data
A catalogue record for this book is available from the British Library

Typeset in Times New Roman
by codeMantra

ISBN 13: 978-0-367-65682-9 (pbk)
ISBN 13: 978-1-138-48921-9 (hbk)

Contents

Citation Information

The chapters in this book were originally published in the journal *Structure and Infrastructure Engineering*, volume 11, issue 4 (April 2015). When citing this material, please use the original page numbering for each article, as follows:

Introduction

Design, assessment, monitoring and maintenance of bridges and infrastructure networks
Fabio Biondini and Dan M. Frangopol
Structure and Infrastructure Engineering, volume 11, issue 4 (April 2015) pp. 413–414

Chapter 1

Reinforced concrete and masonry arch bridges in seismic areas: typical deficiencies and retrofitting strategies
Claudio Modena, Giovanni Tecchio, Carlo Pellegrino, Francesca da Porto,
Marco Donà, Paolo Zampieri and Mariano A. Zanini
Structure and Infrastructure Engineering, volume 11, issue 4 (April 2015) pp. 415–442

Chapter 2

The art of arches
Man-Chung Tang
Structure and Infrastructure Engineering, volume 11, issue 4 (April 2015) pp. 443–449

Chapter 3

Innovative steel bridge girders with tubular flanges
Richard Sause
Structure and Infrastructure Engineering, volume 11, issue 4 (April 2015) pp. 450–465

Chapter 4

Effects of post-failure material behaviour on redundancy factor for design of structural components in nondeterministic systems
Benjin Zhu and Dan M. Frangopol
Structure and Infrastructure Engineering, volume 11, issue 4 (April 2015) pp. 466–485

Chapter 5

Design and construction of two integral bridges for the runway of Milan Malpensa Airport
Pier Giorgio Malerba and Giacomo Comaita
Structure and Infrastructure Engineering, volume 11, issue 4 (April 2015) pp. 486–500

For any permission-related enquiries please visit:
http://www.tandfonline.com/page/help/permissions

Notes on Contributors

Stefania Arangio is a Research Associate in the Department of Civil and Geotechnical Engineering at the University of Rome 'La Sapienza', Italy.

Marco Belloli is a Professor in the Department of Mechanical Engineering at Politecnico di Milano, Italy.

Fabio Biondini is a Professor of Structural Engineering in the Department of Civil and Environmental Engineering at Politecnico di Milano, Italy.

Franco Bontempi is a Professor of Structural Analysis and Design in the Department of Civil and Geotechnical Engineering at the University of Rome 'La Sapienza', Italy.

James M.W. Brownjohn is a Professor of Structural Dynamics in the Vibration Engineering Section in the College of Engineering, Mathematics and Physical Sciences at the University of Exeter, UK.

Giacomo Comaita is a Surveyor at Studio Malerba, Milan, Italy.

Francesca da Porto is an Associate Professor in the Department of Civil, Architecture and Environmental Engineering at the University of Padova, Italy.

Giorgio Diana is an Emeritus Professor in the Department of Mechanical Engineering at Politecnico di Milano, Italy.

Marco Donà is a Research Associate in the Glass & Façade Technology Research Group at Cambridge University, UK.

Dan M. Frangopol is the inaugural holder of the Fazlur R. Khan Endowed Chair of Structural Engineering and Architecture at Lehigh University, Bethlehem, USA.

Bruno Godart is the Deputy Head of the Department of Materials and Structures at IFSTTAR at the Université Paris-Est, Champs-sur-Marne, France.

Sehwan Kim is an Assistant Professor in the Department of Biomedical Engineering at the College of Medicine at Dankook University, Yongin, Republic of Korea.

Ki-Young Koo is a Lecturer in the College of Engineering, Mathematics and Physical Sciences at the University of Exeter, UK.

David List is the General Manager at Tamar Bridge and Torpoint Ferry, Plymouth, UK.

Pier Giorgio Malerba is a Professor in the Department of Civil and Environmental Engineering at Politecnico di Milano, Italy.

Claudio Modena is a Professor Emeritus of Structural Engineering in the Department of Civil, Architecture and Environmental Engineering at the University of Padova, Italy.

Konstantinos G. Papakonstantinou is an Assistant Professor in the Department of Civil and Environmental Engineering at Pennsylvania State University, State College, USA.

Carlo Pellegrino is a Professor in the Department of Civil, Architecture and Environmental Engineering at the University of Padova, Italy.

NOTES ON CONTRIBUTORS

Daniele Rocchi is a Professor in the Department of Mechanical Engineering at Politecnico di Milano, Italy.

Richard Sause is the Joseph T. Stuart Professor and Director of Advanced Technology for Large Structural Systems (ATLSS) in the Department of Civil and Environmental Engineering at the ATLSS Engineering Research Center at Lehigh University, Bethlehem, USA.

Andrew Scullion does Technical Authority Systems Engineering at James Fisher Testing Services, UK.

Masanobu Shinozuka is a Professor in the Department of Civil Engineering and Engineering Mechanics at Columbia University, New York City, USA.

Man-Chung Tang is the Chairman of the Board of T.Y. Lin International, China Engineering & Consulting Co. Ltd., San Francisco, USA.

Giovanni Tecchio is a Civil Structural Engineer at Rete Ferroviaria Italiana, Rome, Italy.

Marco Torbol is an Associate Professor in the School of Urban & Environmental Engineering at the Ulsan National Institute of Science and Technology, Republic of Korea.

Matteo Vergani has been affiliated with R4M Engineering in Milan, Italy.

Paolo Zampieri is completing an internship for his Master of Science thesis, Energy Engineering – Renewables & Environmental Sustainability at LEAP Research Center at the University of Padova, Italy.

Mariano A. Zanini works in the Department of Civil, Architecture and Environmental Engineering at the University of Padova, Italy.

Benjin Zhu is a Structural Engineer at ABS Consulting, Inc., Reading, USA.

INTRODUCTION

Design, assessment, monitoring and maintenance of bridges and infrastructure networks

Fabio Biondini and Dan M. Frangopol

Bridges and infrastructure systems, due to their inherent vulnerability, are at risk from ageing, fatigue and deterioration processes due to aggressive chemical attacks and other physical damage mechanisms. The detrimental effects of these phenomena can lead over time to unsatisfactory structural performance under service loadings or accidental actions and extreme events. The current condition ratings of stocks of existing bridges and infrastructure networks indicate that the economic impact of structural deterioration is exceptionally high and emphasise the importance of maintenance and repair of structurally deficient bridges. To deal with these problems, bridge design criteria and methods need to be revised to account for a proper modelling of the structural system over the bridge life-cycle by taking the effects of deterioration processes, time-variant loadings, maintenance actions and repair interventions into account.

Relevant advances have been accomplished by the scientific community and engineering profession in the design, assessment, monitoring, maintenance and management of sustainable and resilient bridge structures and infrastructures. These advances have been presented and discussed at *The Sixth International Conference on Bridge Maintenance, Safety And Management* (IABMAS 2012), held in Stresa, Italy, from 8 to 11 July 2012 (http://www.iabmas2012.org). The First (IABMAS'02), Second (IABMAS'04), Third (IABMAS'06), Fourth (IABMAS'08) and Fifth (IABMAS 2010) International Conferences on Bridge Maintenance, Safety and Management were held in Barcelona, Spain, 14–17 July 2002, Kyoto, Japan, 18–22 October 2004, Porto, Portugal, 16–19 July 2006, Seoul, Korea, 13–17 July 2008 and Philadelphia, PA, USA, 11–15 July 2010, respectively.

IABMAS 2012 has been organised on behalf of the *International Association for Bridge Maintenance And Safety* (IABMAS) under the auspices of Politecnico di Milano. The objective of IABMAS is to promote international cooperation in the fields of bridge maintenance, safety, management, life-cycle performance and cost for the purpose of enhancing the welfare of society (http://www.iabmas.org). The interest of the international bridge engineering community in the fields covered by IABMAS has been confirmed by the significant response to the IABMAS 2012 call for papers. In fact, over 800 abstracts from about 50 countries were received by the Conference Secretariat, and approximately 70% of them were selected for final publication as technical papers and presentation at the Conference within mini-symposia, special sessions and general sessions, for a total of 555 papers scheduled at IABMAS 2012.

The extended versions of several selected papers presented at IABMAS 2012 and invited papers are published in this special issue of *Structure and Infrastructure Engineering*. These papers provide significant contributions to the process of making more rational decisions in bridge design, assessment, monitoring and maintenance. The paper by Modena et al. presents typical deficiencies and retrofitting strategies of existing reinforced concrete and masonry arch bridges in seismic areas. Tang gives an overview of design concepts and technical issues of modern and ancient arch bridges, with focus on aesthetics, structural shapes and construction materials. Sause presents experimental and numerical results for I-shaped tubular flange girder bridges and shows the advantages of this structural system in comparison with conventional I-girder bridges. Zhu and Frangopol investigate the effects of post-failure material behaviour on redundancy factors for design of structural components in nondeterministic systems, with emphasis on steel highway bridges. Malerba and Comaita present the aeronautical and railway specifications, the design criteria and the construction process of two integral concrete bridges for the runway of Milan Malpensa Airport in Italy. Godart discusses the main pathologies, repair activities and management of old prestressed beam and slab concrete bridges. Biondini and Vergani propose the formulation of a three-dimensional deteriorating beam finite element for damage modelling and nonlinear analysis of concrete structures under corrosion, with application to a reinforced concrete arch bridge. Diana et al. present an overview of wind tunnel activities and methodologies to support the design of long span suspension bridges considering aerodynamic phenomena. Brownjohn et al. describe a range of measurement technologies and applications in structural identification and performance diagnosis of suspension bridges. Arangio and Bontempi investigate the identification of damage and structural health monitoring of a cable-stayed bridge based on Bayesian neural networks. Finally, Shinozuka et al. propose a remote monitoring system for a wide range of structural health-monitoring applications, from single structures to large-scale infrastructure networks.

The guest editors would like to thank the authors and the reviewers for their contributions to this special issue and hope that this collection of papers will represent a valuable reference for scientific research and engineering applications in the fields of design, assessment, monitoring and maintenance of bridges and infrastructure networks.

Reinforced concrete and masonry arch bridges in seismic areas: typical deficiencies and retrofitting strategies

Claudio Modena, Giovanni Tecchio, Carlo Pellegrino, Francesca da Porto, Marco Donà, Paolo Zampieri
and Mariano A. Zanini

In recent years, appraisal of the condition and rehabilitation of existing bridges has become an ongoing problem for bridge owners and administrators in all developed countries. Reliable methodologies are therefore needed in the assessment and retrofit design phases, to identify the vulnerability of each bridge class. The specific problems of common arch bridge types are discussed herein, for both reinforced concrete and masonry structures, proper interventions for their static and seismic retrofitting are illustrated and several examples of applications are provided. Retrofitting is usually coupled with functional refurbishment, according to a methodological approach that takes into account bridge characteristics, state of maintenance and functional requirements, and environmental aspects connected with repair and strengthening systems.

1. Introduction

The performance of existing bridges in most developed countries is inadequate: many of them were designed for performance levels that have become progressively insufficient by increasingly demanding traffic conditions and structural safety requirements. Ever-greater numbers of vehicles, heavier weight/axle loads, higher traffic volumes, increasing vehicle speeds and related dynamic effects are all reflected in the updating process of codes and standards.

In Italy, arched structures are very common in the existing in-service bridges. Most modern masonry arch bridges date back to the period 1850–1930; reinforced concrete (RC) arches were built as pioneering applications after about 1910–1920 and were later frequently used during the post-Second World War reconstruction period, after which RC and prestressed reinforced concrete (PRC) decks with prefabricated girders became more competitive in terms of construction costs than cast-*in situ* RC arches, which required onerous centring.

Existing arch bridges, in most cases underdesigned for such service conditions, not only require wider decks and protection of various types of lanes in order to improve user safety, but also require measures to counteract deterioration and damage propagation effects which undermine the efficiency of all components. These processes, due to environmental (chemical or physical) action, are emphasised in a general context of poor or non-existent maintenance procedures.

In Italy, as in other seismic areas, in addition to the above-mentioned 'vulnerabilities', fragility due to seismic action must also be taken into account when evaluating the vulnerability of arch bridges. Seismic hazard has only recently been recognised by national structural codes (e.g. in Italy, Ordinance of the Presidency of the Council of Ministers No. 3274, 2003). This means that in many cases bridges are extremely vulnerable, as they were designed and constructed without any seismic design specifications at all. What makes this situation particularly critical is that the new generation of Italian structural codes (e.g. Italian Ministry of Infrastructures, 2008) makes seismic verification compulsory, so that proper retrofitting of 'strategic' structures like most bridges must be planned for Civil Protection purposes. Owing to the aforementioned situation, keeping a transportation network efficient is becoming an overwhelming task for public agencies, especially as not enough resources are available to tackle the 'entire problem' in good time, as clearly indicated in different standards (ISO 13822, International Standard, 2010).

Establishing principles for assessing and retrofitting existing structures is essential, as repair and retrofitting are based on an approach that is substantially different from designing structures and requires knowledge beyond the scope of design codes. The ultimate goal is *to limit construction work to a strict minimum*, a goal which clearly matches the principles of sustainable development and follows the guidelines of current standards (e.g. BD

79/06, British Standard, 2006), when such regulations are in force.

In this context, it is crucial:

- To define adequate, reliable assessment procedures, capable of identifying and taking due account of the specific sources of 'vulnerability' of many types of bridges (Carturan, Pellegrino, Rossi, Gastaldi, & Modena, 2013; Carturan, Zanini, Pellegrino, & Modena, 2014), including their seismic vulnerability (Zanini, Pellegrino, Morbin, & Modena, 2013). This lies at the base of unavoidable planning intervention procedures, establishing priorities, allowing for the optimum use of 'limited resources' and of design strategies really capable of devising 'ways for extending the life of structures whilst observing tight cost constraints'.
- To approach the design of 'repair + retrofitting + refurbishment' of existing bridges in a very comprehensive fashion, aiming at providing for all deficiencies, while maximising the contribution to this end of the original design and construction technologies.

Assessing the state of an existing bridge does require a complex overall approach. It may be described as a rational process, consisting of a preliminary *Knowledge Phase* that involves the use of both standard procedures, such as archive research and *in situ* and laboratory tests for the characterisation of material properties, and less conventional tools, such as structural monitoring and dynamic identification techniques. This preliminary phase is followed by the *Analysis Phase* (Figure 1). These two conceptually related phases are not 'one-way'. Feedback of the results of structural analysis must come from reiterative checking of evidence emerging in the Knowledge Phase. In addition to the more commonly addressed problems of static strength and stability, consideration must also be given to a broad range of factors including dynamic and seismic behaviour, long-term deformations, fatigue effects and durability (functional efficiency), and other aspects. When appropriately applied, the presented safety assessment procedure (Steps 1 and 2) represents a sound basis for selecting intervention strategies as well as controlling their efficiency (Zanardo, Pellegrino, Bobisut, & Modena, 2004).

The subsequent choice of the best intervention in terms of materials and techniques (*Third Phase – Design of retrofit intervention*) therefore depends on several factors. The main ones are related to the static scheme of the bridge (simply supported or continuous girder bridge, arch bridge, frame bridge, etc.), to the characteristics of its structural components (deck, piers, abutments, bearings, etc.) and to the type of loads (static, dynamic). Other fundamental aspects to be taken into account when designing the retrofit interventions are the actual state of maintenance of the structure (such as ongoing degradation processes), and the intervention techniques/phases, that should interfere as little as possible with the usual traffic conditions. Requirements of durability and compatibility of materials should be satisfied as well, and non-structural factors such as functional requirements and aesthetic issues should be clearly identified for sustainability of operations.

This paper describes the main degradation processes and typical original design defects of several common kinds of arch bridges, in both concrete and masonry. Rehabilitation operations on the original structure are discussed, making an innovative contribution in providing an overview of several examples rather than focusing on general aspects, and presenting a methodological approach which takes into account the characteristics of the bridge, its state of repair and required maintenance and environmental aspects connected with repair and strengthening systems.

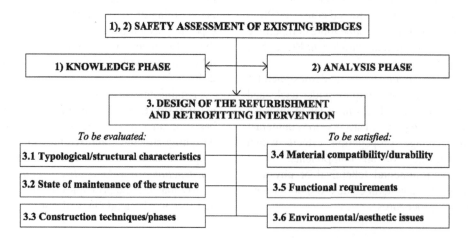

Figure 1. Flowchart of methodology proposed for assessment and retrofitting operations. (1) Knowledge phase, (2) analysis and (3) design of the retrofitting intervention.

2. Assessment and retrofitting of RC arch bridges

In Italy, the most common types of road bridges are in RC and PRC. Most of them were built between 1945 and 1960 in a time of severe shortages of construction materials, and without any motivation for quality control. They include medium to long-span bridges with tied arches, through arches and deck arches, built all over Italy immediately after the Second World War.

The problems of the assessment, rehabilitation and refurbishment of these RC bridge types are discussed here, with specific reference to three representative structures [in Figure 2, the configuration of the bridges before (on the left) and after (on the right) the intervention is shown], which constitute a homogeneous set, all being part of the post-Second World War reconstruction on the Adige (the second longest river in Italy). The deficiencies these structures revealed after 50 years of in-service life are typical, so that this retrofitting description is suitable for

considerations of a general order. A brief description of the structures before retrofitting operations is given below.

Albaredo Bridge. This bridge has six spans for a total length of approx. 230 m (Figure 3). Three spans, on the right bank of the Adige, have an effective length of about 23 m, and are composed of simply supported girder decks. The other spans, about 53 m long, are Nielsen-type arches (tied-arch bridges with inclined hangers). The roadway was originally 6.00 m wide, flanked by two footpaths just over 1 m wide.

Sega Bridge. This bridge has two spans, respectively, 60 and 25.45 m long (Figure 4). The deck, originally 8.34 m wide, is supported by transversal RC portal frames resting on two massive arches. The section of the main arches varies in height (from 1.2 to 1.9 m in the longest span) and is 2.0 m wide.

San Francesco Bridge. This RC bridge is the only one of the case-studies which is located inside the walls of

Figure 2. Panoramic views of bridges adopted as examples, before (left) and after (right) retrofitting. (a, b) Albaredo Bridge, (c, d) Sega Bridge and (e, f) San Francesco Bridge.

Figure 3. Albaredo Bridge, current state. (a) General views and (b) longitudinal section.

Verona, a historic city centre. It is symmetric in plan, with three arched spans, the central one measuring 41.45 m and the side spans 35.81 m (Figure 5). The upper structure consists of a 14-cm-thick RC slab and a 25-cm high grillage of downstanding beams. These are supported on pillars distributed longitudinally and transversely with a pitch of 2.0 m, and varying in height from the crown (0.30 m) to the arch springing (approximately $H = 4.66$ m

in central span). The main structure is the lower RC arch barrel, which varies in thickness from 40 to 70 cm. Thermal joints are provided in the deck, in the two central piers and abutments.

Only strengthening procedures applied to the super-structure are discussed herein. No information on the strengthening of existing foundations is included, since this depends on soil type and condition and foundation typology.

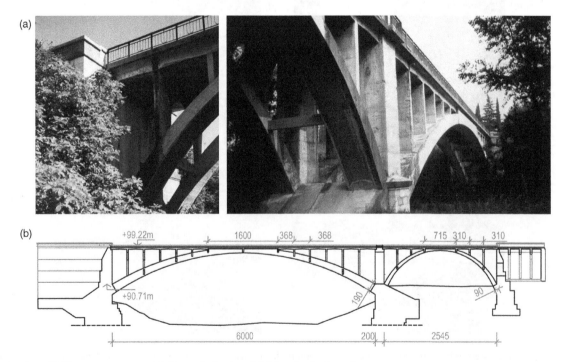

Figure 4. Sega Bridge, current state. (a) General views and (b) longitudinal section.

Figure 5. San Francesco Bridge, current state. (a) General view and (b) longitudinal section.

2.1 Typical deficiencies of RC arch bridges

2.1.1 Degradation

Deterioration processes in RC bridges are due to environmental action, the effects of which are mainly accelerated over time by poor maintenance. Environmental actions are classified, respectively, into chemical, physical and biological degradation mechanisms (SB-ICA, 2007):

- *physical action* is typically due to freeze–thaw cycles, water penetration, thermal variations and environmental vibrations;
- *chemical action* is most frequently due to carbonation of concrete, corrosion of bars (oxidation of metal, causing loss of material), the action of salt, alkali–silica reaction, sulphate attack;
- *biological degradation* is due to accumulation of earth, debris, etc., and the activity of living organisms.

Among physical actions, as well as accidental damage due to impacts and cracking caused by thermal effects, freeze–thaw cycles are one of the most common physical causes of degradation of externally exposed RC elements. These cycles may occur when pore size and distribution in the concrete are inadequate. In combination with critical water saturation (>91%), freezing water may cause concrete to deteriorate, due to volume expansion. Signs of deterioration are continual loss of concrete surface, local pop-outs or micro-cracking and loss of concrete strength at later stages. In general, concrete with a low water–cement ratio has higher frost resistance than concrete with a high ratio. Proper resistance to frost is provided if the air content is approximately 4% in volume and if air-bubbles are well distributed.

Cracking and spalling of the concrete cover, due to carbonation and bar oxidation, are the types of degradation affecting most elements of existing RC bridges, even when they are exposed to non-aggressive environmental conditions (Figures 6–8). The alkaline nature and density of concrete represent a chemical and physical barrier against corrosion. Durability depends on the protection afforded by the surrounding concrete against penetration of chlorides, water and oxygen, which are some of the essential ingredients inducing corrosion of steel reinforcements. In particular, carbonation of the concrete surface, when the concrete cover is limited, leads to corrosion caused by water and oxygen. General corrosion is associated with the formation of iron oxides, commonly called 'brown rust'. The volume of these oxides is several times greater than that of the parent steel. The volumetric expansion of a corroded bar generates tensile hoop strains in the surrounding concrete, leading to longitudinal cracking and subsequent spalling of the concrete cover.

2.1.2 Original design and construction defects

Many deficiencies in existing RC bridges are the consequence of lack of durability rules in the original design and poor quality control during construction, leading to early deterioration of structural performance. The most common design defects of superstructure elements are as follows:

- *Insufficient concrete cover* (widespread in existing RC bridges), offering insufficient protection against

Figure 6. Albaredo Bridge, current state. (a) Carbonation, bar oxidation and concrete spalling; (b) joint defects and (c) cracking and degradation due to inadequate drainage.

carbonation; corrosion then reduces the effective section of reinforcing steel bars (Figures 6–8).

- *Sub-standard quality of concrete*, e.g. insufficient compaction, poor curing, excessive porosity, improper constituents (aggregate, admixtures, water). One related aspect, which may lead to corrosion effects in post-tensioned cables, is *insufficient grouting of tendon ducts*.
- *Insufficient criteria of reinforcement design*, due to poor-quality standards regarding over-strength factors and detailing rules, giving rise to dynamic amplification effects, possible increments in variable axle loads, shrinkage and thermal effects; all these often result in poor confinement of elements and inadequate shear reinforcement.

- *Undersizing of secondary elements for effective traffic loads*, in terms of thickness, stiffness and reinforcement. Secondary elements such as RC upper slabs or cross-beams, the function of which is to distribute local point loads and transfer internal forces to the main structures, are often undersized. This is generally due to underestimation of increases in true axle loads during the in-service life of the structure. The consequences are shear cracks, bending cracks and large deflections, which are often visible to the naked eye.
- *Weakness of details*, which is closely related to the lack of durability rules in the original design. Structural connections, nodes and details (e.g. tie-rod anchorages in arches) without any protection,

Figure 7. Sega Bridge, current state. Carbonation, concrete spalling and corrosion of reinforcement.

Figure 8. San Francesco Bridge, current state. (a) External and (b) internal views of arched structure.

are the components most frequently exposed to environmental agents, representing the onset of degradation. Approximate solutions have frequently been adopted for *in situ* casting during construction, without ensuring the minimum required cover, with overlapping bars not adequately anchored.

- *Lack of (or inefficient) bearings*, e.g. corroded mono-directional steel restraints or hinges which do not ensure thermal movement or static rotation, worn rubber neoprene pads.
- *Lack of (or insufficiently durable) expansion joints*, worn away due to poor maintenance and dynamic effects due to the passage of vehicles. Expansion joints are exposed to weathering, and water passing through them is the main source of degradation of RC elements.
- *Inadequate waterproofing and drainage systems*, the major cause of deterioration of concrete bridge components when de-icing chemicals are used.
- *Lack of seismic design specifications*: earthquake-resistant rules for bridge design have only recently been adopted by the Italian code (but more in general also in European earthquake-prone areas; Modena, Franchetti, & Grendene, 2004). Thus, the piers and abutments of existing bridges are not strong enough to resist lateral seismic forces. The deficiencies mainly concern shear reinforcement, detailing for section ductility and foundation capacity. Bearings and supports are also generally inadequate for the transmission of inertial loads to the substructure.

In the case histories reported here as examples, all the aforementioned aspects can be shown.

Albaredo Bridge, current state. In the three main arched spans, structural studies and tests demonstrated that the existing arch did not require any strengthening work, as the main strut of the arch had a quite generous section (T-section, 1.12 m high, 0.6 m wide). However, the tie-rods were quite badly over-stressed, and the durability of the reinforcement was not guaranteed, since it was only protected by a thin concrete cover with extensive cracking in some areas (Figure 6). The concrete slab showed marked deficiencies, due to inadequate thickness (25 cm) and excessive deformability of the cross-beams, which triggered longitudinal bending stresses higher than those corresponding to the design scheme, in which the cross-beams acted as restrainers. Both the main beams and the upper RC slab/cross-beam system of the secondary girder spans were inadequate.

Sega Bridge, current state. The bridge revealed substantial structural deficiencies, mainly in secondary members (Figure 7). The main arches and columns did have robust concrete sections, suitable for supporting current loads, but the transverse reinforcement of the arches was insufficient to stabilise the longitudinal bars. The deck was affected by severe deterioration, due to carbonation and localised rainwater damage (no waterproofing). The low percentage of longitudinal reinforcement of the RC deck sections caused stresses not compatible with the low strength of the smooth steel bars (FeB22k type, $f_{yk} = 215$ MPa). In addition, as the existing deck intrados was curved, the lack of transverse reinforcement led to inadequate prevention of delaminating effects in the curved longitudinal bars.

San Francesco Bridge, current state. Serious deterioration effects were observed in all structural members, affected by carbonation, corrosion of reinforcements and spalling of concrete (Figure 8). The inner columns supporting the upper RC slab were built with very poor quality materials, and were characterised by insufficient axial load capacity. The carbonation on these internal pillars reached deeper than the concrete cover, to about 62 mm, and biological degradation was also observed, with mud and debris settling in the lower part of the vaults, near the arch springing. The RC arch vaults did not show particular static deficiencies, although widespread deterioration of materials was observed, with a carbonation depth of 32 mm. The components of the deck, upper RC slab and grillage of downstanding beams were found to be seriously undersized to bear current traffic loads for a first-category bridge. Serious local damage affected the expansion joints of each span, due to the deficient drainage system.

2.2 Strategies for retrofitting RC arch bridges: innovative versus traditional

Static rehabilitation, to bring safety conditions up to current standards for in-service loads, is generally the starting point of structural interventions on existing bridges. The live-load capacity of bridges must be restored according to the type of structure and its current state. The type of intervention must be designed not only with reference to technique but also to materials.

The choice of appropriate materials greatly influences not only costs but also operating mode and future durability. It is essential to examine the physical and chemical compatibility of both new materials and existing ones, and to choose materials that ensure the best performance in terms of durability over time, depending on the environmental conditions in which they are to operate. Examples of frequently used compatible materials are *stainless steel* for insertion of new exposed reinforcement bars (or at least steel carbon galvanised bars, which are cheaper); *composite materials [Fiber-Reinforced Polymer (FRP)]* for flexural/shear strengthening and confinement of RC members; *lightweight concrete*

for substitutive or additional RC deck slabs and *thixotropic shrinkage-compensated mortar* for renewing concrete covers.

2.2.1 Rehabilitation and treatment of deteriorated surfaces

Deteriorated materials must generally be systematically treated to arrest degradation. The most seriously damaged parts of concrete covers should generally be hydro-demolished. More moderate treatments by sand-blasting can be used for well-preserved concrete, in which carbonation has not penetrated deeper than a few mm. These operations can be done mechanically over large surfaces (like deck slabs) and manually for smaller elements.

The entire surface area is then treated by sand-blasting, until clean degreased surfaces are obtained, with no fine particles which could affect good adherence of subsequent plastering. All exposed rebars are sanded down to white metal, blown with pressurised air jets and treated with an anti-corrosive agent. After integration of the most corroded bars if necessary, new plaster is applied to the cover with thixotropic shrinkage-compensated cement mortar, FRPs (Pellegrino, da Porto, & Modena, 2011). These phases were applied for rehabilitation of the degraded surfaces of all the bridges presented here, although to different extents.

In the Albaredo Bridge (Figure 9), the surface at the extrados of the deck was mechanically hydro-demolished, thus preparing it for the subsequent casting of a supplementary RC slab. In all linear elements (main struts, tie-rods and lateral tie-beams), the operations for the treatment of the degraded surfaces were executed manually, and the reconstruction cycle of the carbonated concrete cover and integration of bars was completed as described earlier.

In the San Francesco Bridge (Figure 10), the main operational phases on the existing arches and spandrel walls involved: manual removal of mud and debris from

Figure 9.   Albaredo Bridge. (a, b) Surfaces of main elements hydro-demolished before reconstruction of concrete cover and (c) casting of additional RC slab.

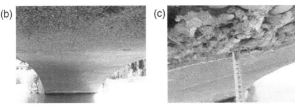

Figure 10. San Francesco Bridge. Hydro-demolition of arches: (a) lateral view, (b) view from intrados and (c) carbonated cover removal.

inner structure (floodwater entering through openings in piers); cleaning of concrete surfaces to remove laitance, dust and dirt by sand-blasting; manual hydro-demolition of the degraded concrete cover to an average depth of 20 mm, and application of new structural plaster with suitable rheoplastic shrinkage-compensated mortar and a final protective coating of arch intrados and spandrel walls.

2.2.2 Static retrofit

There are several procedures for static strengthening of RC bridge members, all generally aiming at increasing the live-load capacity. These solutions should respect the essential role of the old structure, making it a constitutive part of the new layout, through use of compatible materials and possibly removable intervention techniques. Strengthening may be limited to a local increase of transfer mechanisms in secondary elements or, in primary members, aimed at improving the overall resistance of the structure. A list of usual retrofit interventions for RC arch bridge superstructures is given below.

- *Dead-load reduction*: this may be achieved by removing part (generally the upper RC slab) or the whole existing heavy concrete deck, and replacing it by a lighter structure (e.g. a lightweight concrete slab in the first case, or a steel or composite deck for complete substitution).
- Increased flexural and shear capacity of RC members, with reconstruction of the concrete cover: this is generally coupled with integration of reinforcement bars, which usually implies:

   (i) installation of new small-diameter longitudinal bars with improved adhesion, which must be positioned next to the oxidised bars (bar diameter must be compatible with new cover thickness; this operation is generally preceded by treatment of the deteriorated surface);
   (ii) addition of stirrups to increase the shear strength of transversal or longitudinal beams, to support added longitudinal bars and better confine member sections (e.g. in cantilever or frame piers);
   (iii) addition of local reinforcement to enhance behaviour with respect to local dynamic effects.

- *Installation of a post-tensioning system*: this may be internal or external, and is highly versatile. It can be used to relieve original tension over-stresses associated with axial load in truss members or flexure or torsion in box or girder decks; to reduce shear effects, if vertically inclined and polygonal tendons are used; to develop additional bridge continuity and/or reduce undesirable displacements or global deflections of members.
- *FRP strengthening*: adding composite materials in sheets, rods, strips or plates to compensate the reinforcement deficiency in over-stressed elements has become widespread in concrete bridge strengthening. FRPs are frequently carbon (CFRP), glass or aramid polymers, and are bonded in sheets or strips to the tension face (Pellegrino & Modena, 2009) or wrapped around the section of the element (Pellegrino & Modena, 2010) in the exact location where increased flexural or shear strength is required (Pellegrino & Vasic, 2013). The system takes up minimal space, is easy to handle and install, and the material has high tensile strength, no corrosion problems and excellent fatigue properties.
- *Local demolition and complete reconstruction* of underdesigned members with the same materials, characterised by higher performance and durability.
- *Creation of dual resisting systems*, acting in parallel for partial load transfer to the new structural system: these are often obtained by adding new steel or composite structures.
- *Substitution of bearings*: the inefficiency or even the complete lack of existing bearings for tie-rod arch bridges may require a new support system, to allow for expansions caused by temperature variations and easy access for inspections. New uni- and multi-directional bearings, positioned under each span, are often in steel, with teflon-treated surfaces [Polytetrafluoroethylene (PTFE)].
- *Elimination of deck joints*: expansion joints are typically one of the most critical points in existing structures, and are important for the durability and

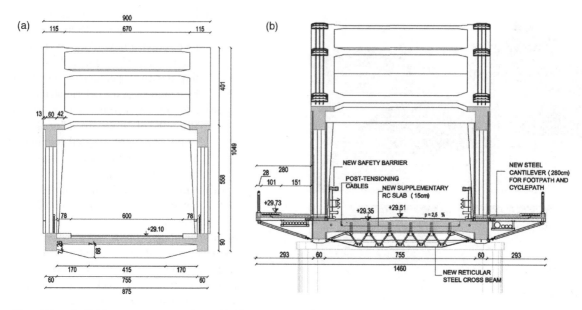

Figure 11. Albaredo Bridge, transverse sections. (a) Current state and (b) after retrofitting with new reticular steel cross-beams and lateral cantilevers.

riding quality of a bridge. One operation of essential importance, both statically and functionally, involves eliminating deck joints, if there are not unfavourable static consequences for the structure. This operation allows the entire deck to be the consolidated in terms of transmission of horizontal forces, without significantly altering its original static behaviour under vertical loads or changing the distribution of bending moments or reactions at piers.

These retrofit design solutions and techniques are now described in detail, with reference to their application in the RC arch bridges adopted as examples:

Albaredo Bridge, retrofit intervention. The general strengthening approach is shown in Figure 11. The current

state of the bridge indicated a retrofit design option that might preserve the main components of the existing bridge, i.e. the system of arches and longitudinal beams. The original design defects were solved by:

- addition of a system of steel cross-beams and tie-rods (Figures 12 and 13), parallel to the existing members and respecting the original geometry;
- supplementary cover of the existing concrete slab (Figure 14) and relative longitudinal prestressing.

The new tie-rods of the main RC arches were placed vertically at regular intervals, at the mid-span of the anchorages of the existing inclined rods (Figure 12) and tensioned with suitable prestressing. The groups of four rods are in stainless steel, anchored to the main arch by external plates connected to suspenders allowing a

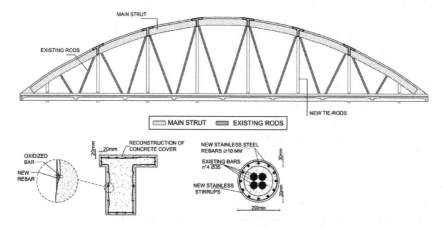

Figure 12. Albaredo Bridge. Rehabilitation and strengthening, with insertion of new longitudinal bars and stirrups on struts, and additional tie-rods for the main arches.

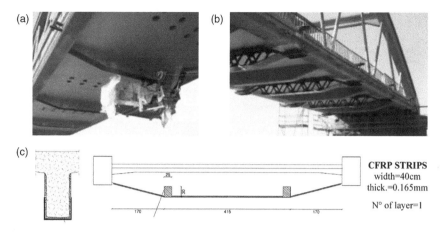

Figure 13. Albaredo Bridge. (a) Application of FRP strips at intrados of existing RC cross beams, (b) new tapered steel cross-beams and (c) properties of CRFP strips.

Figure 14. Albaredo Bridge. Realisation phases of the new supplementary post-tensioned lightweight concrete slab.

self-balancing system of stress transmission. The following effects were thus obtained:

- reduction of tension in existing rods;
- reduction of bending stress in longitudinal beams;
- efficient support for the new steel cross-beams, which have the same tapered shape as the existing RC cross-beams, and are hung on the new vertical tie-rods;
- adequate safety conditions of the entire bridge, thanks to the size of the new suspenders, which can resist even if one of the existing ones should collapse;
- the new reticular steel cross-beams were positioned after imposing a precamber granting the desired reduction of bending in the existing concrete slab;
- the existing cross-beams were reinforced with the reconstructed concrete cover and application of CFRP strips at the intrados (Figure 13). Work was completed with a supplementary lightweight concrete slab 15 cm thick, with a longitudinal post-tensioning system (Figure 14). The cables were simply laid in conduits on the upper surface of the existing concrete slab and incorporated in the added RC slab. Longitudinal post-tensioning ensures efficient transmission of a significant amount of the tensile stress transmitted by the arches, thus making the existing longitudinal beams reliable.

When the continuous slab covered the whole system of spans, the joints between adjacent decks were eliminated. Expansion joints were only placed at the ends of the bridge, to allow temperature expansion. New multi-directional bearings were positioned beneath each span (Figure 15), with PTFE devices sized to guarantee longitudinal creep proportional to their distance from the abutment on the left bank (fixed), to which the whole deck is longitudinally anchored.

(a)  (b)

Figure 15. Albaredo Bridge. Raising of deck with hydraulic jacks and insertion of PTFE bearings.

Both the main beams and concrete slab cross-beams of the three secondary girder spans external to the arches were inadequate, and required complete reconstruction of the deck (Figure 16). A new composite structure was adopted, composed of four main welded I-beams and an upper RC slab cast on a corrugated metal sheet. The consequent reduction of dead loads and the resulting seismic protection system meant that substantial work on the elevations and foundations of piers could be avoided, except for rehabilitation of carbonated surfaces (concrete cover reconstruction).

Sega Bridge, retrofit intervention. This was differentiated into primary resisting members (RC arches and

(a)

(b)

Figure 16. Albaredo Bridge. Complete reconstruction of the shorter girder spans with a composite deck: (a) lateral view and (b) from intrados.

pillars) and secondary structures (upper RC slab). Except for carbonated surfaces, the main structural members did not require other strengthening interventions, except improved lateral confinement of longitudinal rebars. The usual restoration cycles were carried out on degraded surfaces, with manual hydro-demolition of the cover, treatment of oxidised bars with corrosion inhibitors and restoration of the cover with rheoplastic shrinkage-compensated mortar. The existing reinforcements were also integrated with small-diameter galvanised carbon steel bars.

In detail, transversal bars and open stirrups were added, properly anchored to the concrete section with epoxy resin, and local integrations of steel rebars were performed in correspondence of the most highly oxidised longitudinal ones. Conversely, the deck required massive interventions (Figures 17–19). At the extrados, after removal of the deteriorated concrete cover, the existing RC slab (28 cm thick) was reinforced with longitudinal and transverse elements and an extra concrete slab was cast (average thickness about 8 cm). A transverse post-tensioning system with high-strength bars ϕ 26.5 mm was added to contrast excessive deflections of the new lateral RC cantilevers. Transversal bars were also added at the intrados, a particularly weak point being the lack of transversal reinforcements supporting the longitudinal bars. The curvature at the intrados was eliminated by adding several layers of high-performance shrinkage-compensated mortar manually, to yield a horizontal surface.

San Francesco Bridge, retrofit intervention. Static retrofitting (see Figure 20) involved complete reconstruction of the existing RC upper slab and beam grillage, which was insufficient to bear the traffic loads required for a first-category bridge (before repairs, there was a restriction on heavy trucks). Preliminary studies and analyses showed that the main arch vaults and foundations did not have severe deficiencies, even when the slight increase in the weight of the superstructure (<15% of the overall load) was added, due to essential deck widening (total increase in width of 5.50 m). Reconstruction of the superstructure involved:

- diamond wire cutting of the lateral cantilever and its removal (Figure 21);
- reconstruction of the carbonated concrete of the external surfaces (intrados of arches, lateral spandrel walls) after the above treatment cycles;
- demolition of the upper RC slab of the internal spans;
- removal of mud and debris from inner parts of the bridge;
- manual demolition of inner columns, preserving the starter bars at the base;
- building of foundations for new steel elements, supporting lateral deck widening;

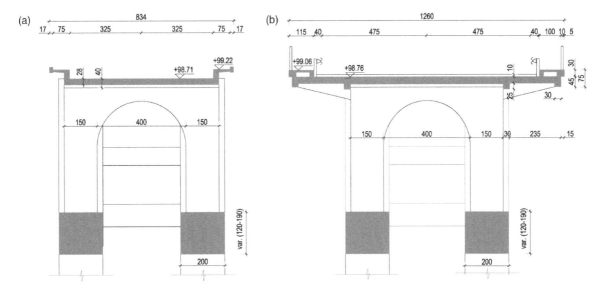

Figure 17. Sega Bridge, transverse sections. (a) Current state and (b) after repair.

- installation of new tilted steel arches ($\phi = 457$ mm, 22 mm thick) and vertical tubular supports (CHS 168, 8 mm thick);
- laying of predalles RC slabs for lateral cantilevers;
- positioning of formwork for inner columns, laying of new reinforcements and casting of new RC pillars (30×30 cm, variable height);
- construction of 30-cm transversal seismic RC walls at the piers;
- construction of upper (25 cm thick) RC deck slab with shrinkage-compensated concrete.

Although a new RC slab was laid down over all three spans, expansion joints could not be eliminated. If deck continuity was completed, additional thrusts would arise in the existing arches (e.g. owing to constrained thermal action), due to the continuity of the upper slab with the arch vault at the crown. As the existing arch vault had not been designed for this extra action, new finger joints were emplaced at the piers, together with new shear connections with stainless steel bars across the joints (Figure 22).

Organising all these construction phases was particularly demanding, as the bridge is one of the most important ones in the city centre. Work was almost always carried out with one traffic lane open (alternating one-way system), on half of the transverse section. The bridge was closed completely only at night or for some hours on Sundays, for launching of the steel arches.

2.2.3 Seismic retrofit

Most existing RC arch bridges are not designed to withstand seismic action. On one hand, their vulnerability is related to the level of seismic action predicted and, on the other, to the intrinsic fragility of components resisting lateral forces. Regarding the seismic response, decks are

Figure 18. Sega Bridge, retrofitting operations. (a) Addition of transverse reinforcement bars at intrados and (b, c) insertion of transverse post-tensioning system at deck extrados.

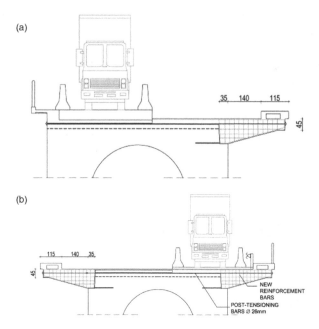

Figure 19. Sega Bridge, construction phases. (a) New lateral cantilever on upstream side and (b) completion on downstream side and inversion of alternating one-way traffic lane.

generally very stiff, resistant to in-plane actions, and their behaviour is elastic. One important aspect is self-weight: the use of lightweight materials (e.g. steel decks) contributes to reducing inertial forces transferred to substructure components, which are the weakest points.

Particularly for tied-arch bridges, generally designed as self-equilibrated decks simply supported on piers and abutments, seismic design deficiencies are identified as follows:

- insufficient bearing capacity;
- undersizing of deck-supporting elements (cap beams, saddles) and joints in relation to differential displacements required on pier tops;
- low shear capacity and low confinement of piers, due to insufficient transverse reinforcements at critical sections;
- insufficient flexural capacity of piers, due to low reinforcement steel ratio and insufficient anchorage of longitudinal bars at the pier bases, leading to possible lap-splice and unbonding effects in longitudinal bars (which are often smooth);
- insufficient shear and rotational capacity of foundations.

In the case of deck arch and through-arch bridges, the deck is integrally connected with the lower arches or barrel vaults (which have great lateral stiffness) at the arch crown. The weakest point for the transfer of lateral forces from the deck to the foundations is generally found in the sections of the arch springings.

Two main design approaches are used for seismic retrofitting, to comply with current code requirements:

- *Strengthening of all weak members* of the substructure and foundations, to increase the capacity and global ductility of the structure, if advantage can be taken of inelastic deformation and hysteretic dissipation in RC members.
- *Passive protection* of the structure by isolation and/or increased damping. The extra damping or isolation system can be calibrated to limit inertial shear forces transmitted to piers and abutments to values comparable with static forces (wind, braking action, etc.). Thus, substructure elements are not required to behave inelastically.

Many techniques have been proposed for seismic strengthening interventions on RC substructure elements (see, among others, Priestley 1995). The most common are:

- enlargement of pier sections, with insertion of longitudinal steel bars and transverse reinforcements;
- additional confinement of piers by steel jacketing;
- use of FRPs for shear and flexural strengthening of columns and cap beams in frame piers;
- addition of local reinforcements to enhance behaviour due to local dynamic effects (in terminal transition slabs, etc.);
- insertion of infill RC walls or new bracing systems between columns in frame piers to increase pier capacity transversely and
- strengthening of foundations: enlargement of plinths, insertion of new piles, jet grouting consolidation, etc.

All these techniques are required to intervene substantially on piers and foundations, although much extra work is often required, with a huge impact on budgets and time schedules for seismic retrofitting. A more economical solution, compatible with the original design, is seismic isolation and/or damping devices. The work involved is closely related with static strengthening because bearings must often be replaced for proper rehabilitation of the superstructure support system, so that coupling seismic devices to new bearings generally requires only a little extra effort.

In the transverse direction, the maximum design force of an additional damping (or isolation) system is chosen to be lower than the capacity of piers and foundations, so that there is no need to strengthen the substructure. In the longitudinal direction, seismic retrofitting involves removing expansion joints and building a continuous deck slab (also for enhanced durability and comfort). These operations are usually carried out to create a kinematic chain system and transfer all longitudinal inertial forces to

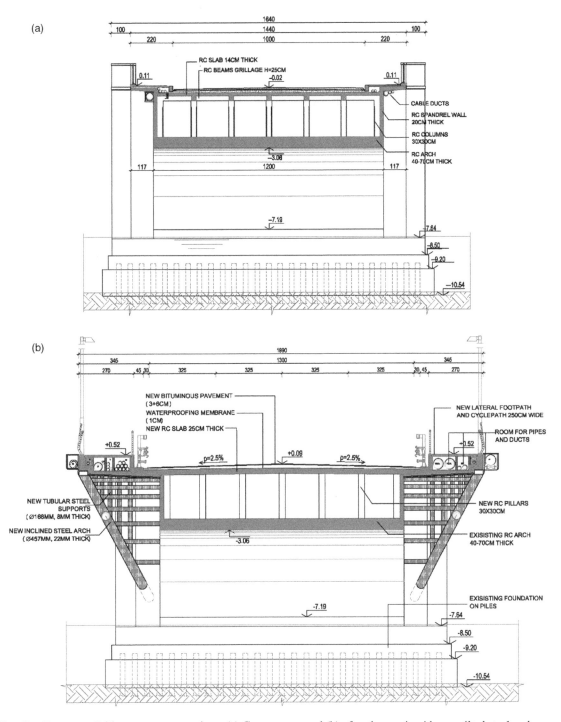

Figure 20.   San Francesco Bridge, transverse sections. (a) Current state and (b) after the repair with new tilted steel arches.

one or more fixed points, where proper devices (e.g. viscous dampers) can later be installed to limit the maximum action transmitted to the substructure if necessary. Piers often do not have sufficient capacity to withstand horizontal longitudinal forces, so that fixed longitudinal restraints are emplaced only at the abutments, after preliminary strengthening of pier walls and foundations.

The aforementioned concepts are illustrated with reference to the cases examined here:

Albaredo Bridge, seismic retrofit. A continuous supplementary RC slab and a palisade with tilted micro-piles were emplaced at the abutment on the left bank of the river to absorb all longitudinal forces due to seismic action. An additional system with elastoplastic dampers (hysteretic steel devices) coupled with PTFE bearings was

Figure 21. San Francesco Bridge. Reconstruction phases of superstructure.

installed on each pier, to limit transversal inertial forces transmitted by the superstructure (Figure 23).

San Francesco Bridge, seismic retrofit. This is one example of the *strengthening approach* to increase the capacity of existing structural parts resistant to horizontal forces. The existing slender pillars were replaced near the arch springings with two suitably reinforced transversal RC walls 30 cm thick (Figure 24). The seismic walls were integrally connected to the new upper RC slab to ensure transfer of inertial forces from deck to foundations where efficient anchorages were emplaced.

2.2.4 Functional refurbishment

Managing authorities frequently ask for bridges to be widened, not only to extend the roadway and add more traffic lanes, but also to create cycle paths and footpaths raised above the roadway and separated from it by safety

Figure 22. San Francesco Bridge. (a) Installation of finger expansion joints and (b) casting of new upper RC slab.

barriers. In most cases, functional refurbishment consists first of adapting the roadway by widening the deck, and then elimination of expansion joints, insertion of approach slabs (to reduce local settlement and provide a gradual transition between roadway and bridge deck) and adjustment of access ramps.

Other finishing works are:

- waterproofing the structure by laying cement or bituminous membranes to protect the roadway (Figure 25);
- building of drains to collect rainwater;
- laying cables and conduits for electricity, gas and water supply facilities;
- inserting safety barriers and parapets, and adapting road signs and
- installing lighting systems.

Structures supporting lateral widening can easily be used to hang pipes and conduits at the intrados, and are usually the most important elements defining the aesthetics of the renovated bridge. There are many design choices, according to materials used, original characteristics and environmental integration of new structures. The solutions adopted for the cases examined here are listed below:

Albaredo Bridge, functional refurbishment. Deck widening (from about 8 to 14.60 m) was achieved with reticular steel structures supporting the external cantile-

Figure 23. Albaredo Bridge. Palisade created at abutment to absorb longitudinal seismic action. Coupling of elasto-plastic dampers with PTFE bearings in the transverse direction.

vers, composed of corrugated metal sheets and a supplementary slab in lightweight RC. The new steel structures were connected to the steel cross-beams, forming an overall system completely integrated with the existing bridge. Widening was sufficient to accommodate a 1.50-m-wide cycle path and a sidewalk of 1.00 m (Figure 26).

Sega Bridge, functional refurbishment. In this case, taking into account the specific construction and geometric properties of the bridge, widening was achieved with post-tensioned RC cantilevers, for a total deck width of 12.60 m (originally 8.34 m). The transverse post-tensioning system limited deflections and crack widths at the extrados of the cantilever structures, thus enhancing durability.

San Francesco Bridge, functional refurbishment. Functional adaptation consisted in the requirement for four traffic lanes instead of three, wider lateral cycle paths, and ample space for pipes and conduits, including high-pressure gas and district heating supplies for residential use. These requirements were particularly demanding and involved widening of 5.50 m of the road platform. Tilted

Figure 24. San Francesco Bridge. Insertion of new transverse RC walls at piers.

Figure 25. Finishing works. (a) Waterproofing of structure, (b) insertion of safety barriers and (c) cables and conduit laying.

(a)

(b)

Figure 26. Albaredo Bridge. (a) View from intrados of widened roading and (b) new lanes for pedestrians and cyclists.

(a)

(b)

Figure 27. San Francesco Bridge. Views of tilted steel arches supporting the lateral widening: (a) from intrados and (b) lateral view.

steel arches were installed, with vertical pillars having the same spacing as the existing internal pillars (2.0 m). This gave a new frontal prospect to the bridge, which now appears as a sort of internal section of the old one, but with new materials.

The use of lightweight materials allowed proper environmental integration of the new structures (Figure 27).

3. Assessment and retrofitting of masonry arch bridges

Most modern masonry arch bridges are part of the historical heritage of the nineteenth century. Due to their long in-service life, often exceeding 100 years, damage was inevitable, partly due to natural weathering and ageing of poorly maintained materials, and partly to increased traffic over time.

Many rehabilitation techniques, derived from the field of historical heritage restoration and reserved in the past for monumental buildings, can also be effectively used for these types of structures. In this context, new concepts are entering structural design practice, supported by guidelines and code standards (e.g. ICOMOS, 2003) which

clearly modify the design approach, the aim being to reduce the impact of over-conservative rules for static and seismic retrofitting, and to minimise interventions as much as possible. The main concepts are:

- assessment of mechanical properties not based on real statistical evaluation, since only limited data from on-site and laboratory tests can be obtained. The estimation can be enhanced using Bayesian updating (once prior statistical distributions are known);
- safety evaluations, based purely on considerations of equilibrium;
- qualitative evaluation of structural performance (observational approach: existing structures as model of themselves) and
- differentiation of safety verification level (limited improvement vs. retrofitting and full compliance with seismic requirements for new structures).

Masonry arch bridges are usually quite robust structural systems (Heyman, 1966, 1972). Their possible weaknesses are not related to the state of stress under permanent uniform symmetric loads, which is typically low with respect to material characteristics (strength), but rather to triggering of antimetric collapse mechanisms due to concentrated vertical traffic loads or horizontal (seismic) forces (Figure 28) (da Porto, Franchetti, Grendene, Valluzzi, & Modena, 2007). In this sense, arch failure may be viewed as a matter of limit equilibrium (mainly depending on geometric characteristics; Gilbert & Melbourne, 1994) activated when external forces create a sufficient number of non-dissipative hinges transforming the structure into a mechanism (Clemente, 1998; Clemente, Occhiuzzi, & Raithel, 1995).

Retrofitting techniques should therefore aim to counteract possible non-symmetric collapse mechanisms due to increased traffic loads (in-service static behaviour) and seismic inertial forces (exceptional conditions). Restoration and strengthening may involve traditional or innovative techniques, according to characteristics (single- or multi-span structures, squat/high abutments, very stiff or slender piers, depressed or semi-circular arches, etc.). The retrofit design solution should also use compatible materials and respect the structural role of the old arch masonry, making it a constitutive part of the new structural layout (Tecchio, da Porto, Zampieri, Modena, & Bettio, 2012).

3.1 Typical deficiencies in masonry arch bridges

The main deficiencies in masonry arch bridges are broadly classified as damage to foundations and to superstructure. The most common defects in foundations include local undermining, differential settlements and masonry dislocations due to loss of mortar joints. The main problem in identifying foundation damage is the difficulty of inspecting underground structures. Therefore, the first step in detecting problems in faulty foundation systems implies observation and analysis of how the superstructure behaves, i.e. the consequence of rotational or differential movements at foundation level. Due to their high stiffness and brittle structural behaviour, masonry bridges cannot generally absorb foundation settlements without structural damage.

Superstructure defects (Figure 29) are easier to detect by visual inspection. The main deficiencies are:

- deterioration of materials, such as degradation and loss of bricks, loss of mortar joints and salt efflorescence in bricks, all often due to inadequate rainwater drainage, freeze–thaw cycles and penetrating vegetation;
- arch barrel deformations, with longitudinal or transverse cracking; opening of arch joints and separation between brick rings in multi-barrel vaults;
- spandrel wall movements: sliding, bulging or detachment from the barrel. Spandrel walls have little inertia and are generally weak elements with respect to out-of-plane behaviour (pressures orthogonal to spandrel walls are due not only to the weight of infill and traffic but also to horizontal transverse seismic action);
- fractures in piers and wing walls; cracking.

As regards RC arch bridges, the typical deficiencies and specific problems for repair and strengthening of

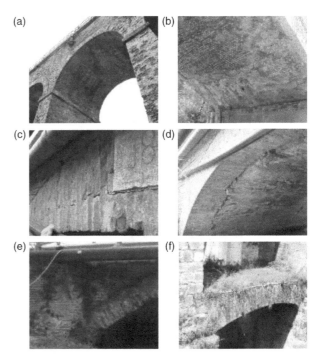

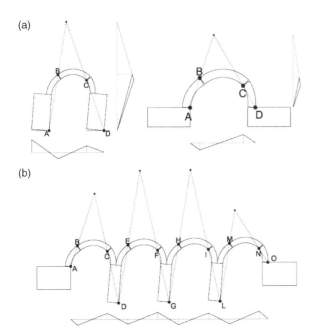

Figure 28. Antimetric collapse mechanisms for masonry arch bridges. (a) Single-span and (b) multi-span bridges.

Figure 29. Typical defects of masonry arch bridges: (a) loss of bricks, (b) salt efflorescence in bricks, (c) opening of arch joints, (d) longitudinal cracking in barrel vault and (e, f) penetration by vegetation.

Figure 30. Sandro Gallo Bridge, current state.

ordinary masonry arch road bridges are presented below, with reference to some representative case histories. A brief description of the structures before retrofitting is given, showing the main deficiencies of their current state:

Sandro Gallo Bridge, current state. The bridge crosses the Excelsior Canal on the Lido of Venice. The bridge was built in the last years of the nineteenth century and the early decades of the twentieth century, with a substantially homogeneous structural arrangement. The bridge has a single-span brick arch (Figure 30), about 0.36 m thick (three brick layers) in the central part, and 0.55 m (four layers) in the side close to the springing. The abutments are made of brick/trachyte masonry with poor-quality mortar, from -1.00 to -5.90 m below the road surface in the older part, and mainly by a massive concrete structure (2.70 m thick) in the more recent part, corresponding to the widening of the bridge.

Structural study involved taking three core samples, and three single and one double flat jack tests on the arch structure. Two more core samples were taken vertically from the abutments on both sides. Results from preliminary tests defined arch morphology: the core samples showed the thickness of the arch at the crown (0.36 m) at 1.27 and 0.53 m from the abutments (0.47 and 0.55 m, respectively). Flat jack tests were performed at three points of the arch (at 1.00, 1.70 and 1.90 m from each abutment). Results showed a moderate state of stress at all points tested (0.25, 0.24 and 0.25 MPa, respectively) and the masonry showed a fairly good response in terms of compressive strength (2.00 MPa).

The core samples from the abutments served to characterise the morphology of the underlying structures and soil stratigraphy. They showed the presence of a 1.00–1.60-m layer of gravel, sand and cobblestone filling under the road surface, and then a difference in terms of foundation characteristics and materials was found, as previously described. The soil was examined as deep as 15.00 m below the starting level of the foundations and the core samples showed a sequence of silty sand and clayey silt. The bridge did not appear to be severely damaged, except for the loss of mortar joints. Preliminary static analysis showed limited bearing capacity with respect to current loads for a first-category road bridge.

Rio Moline Bridge, current state. This is a two-span road bridge with squat piers (Figure 31) built in the

Figure 31. *Rio Moline Bridge*, current state. (a) Panoramic view and (b) details.

eighteenth century over the river Moline in the province of Trento (northern Italy). It has two arches of different lengths (about 6.80 m for the span on the orographic right side and 7.50 m on the left), with one central pier standing in the middle of the river bed. The arches are slightly tilted, with stone voussoirs of variable thickness (40–45 cm); the central pier is 1.70 m wide, and the abutments stretch on each side with wing walls 4.50 m long, making the total length of the bridge about 24 m. The single traffic lane, alternating one-way, has a total width of 3.90 m. Due to these geometric constraints, the bridge can only be used by light traffic and is classified as third category (live load = 5.0 kN/m²).

The bridge appeared to be in very poor condition, and provisional wooden shoring with steel ties had already been put in place. *In situ* tests were made to identify the stratigraphic layers of the flooring and filling; continuous inclined and horizontal cores and endoscopic probes were performed at the springing and in the pier and abutment elevation walls. The masonry was significantly affected by loss of mortar joints, erosion of which had progressively led to loss of connection between the stone blocks and some of them had become loose. The intrados surfaces were also partially damaged by vegetation, with roots penetrating deeply between the joints. In the stone vault structure, discontinuities and cavities were found, with filling characterised by loose material with traces of earth. The abutments and central pier have walls made of an external curtain of larger blocks 45–55 cm thick, and an internal core in dry stone masonry, with decimetric-sized stones.

Gresal Bridge, current state. This bridge is in the province of Belluno, in north-east Italy. It was built in the nineteenth century, is currently used as a road bridge and is an important regional crossing over the Gresal river. The structure is a three-span stone masonry arch bridge (Figure 32), total length 67.40 m: the three spans are almost equal, the single arch clear length being about 15 m. The spans are almost semicircular in shape, with a radius of 7.39 m, slightly increasing at the springing.

The average thickness of the arch crown is 0.60 m, and the maximum height of the two piers, which taper between bottom and top, is 12.75 m. The rectangular-section piers measure 3.50 × 6.99 m at the base (the larger one being orthogonal to the bridge axis). The roadway is 6.09 m wide, and the spandrel stone walls emerge laterally beyond the deck level, forming two 45-cm-thick parapets.

Structural study consisted of three core samples taken from the stone arch and a geometric survey. Two vertical core samples were taken, one from the central part of the arch and one from the pier; a third was drilled in the abutment, on an inclined plane. These studies determined the thickness of the bricks, stone and layers of material between pavement and masonry vault, and characterised the mechanical properties of the infill.

The infill was 0.75 m thick at the arch crown and was composed of loose material, mostly stones and pebbles, with good mechanical characteristics. Visual inspection and structural studies did not show evident structural damage, except for insufficient waterproofing of the masonry, probably due to humidity. Preliminary seismic analysis showed great vulnerability to seismic action, mostly due to the slenderness of the high piers.

3.2 Rehabilitation and retrofit strategies for masonry arch bridges: innovative versus traditional

Two main general approaches can be identified for retrofitting masonry arches:

- strengthening, to recover and increase the load-bearing capacity of the original structure (by improving material properties and connections, thickening the old structure with the same materials, etc.);
- creation of resistant systems acting in parallel with the old structure or directly increasing the strength of original members (e.g. by adding tensile reinforcements in the original masonry section).

(a)   (b)

Figure 32. Gresal Bridge. (a) Panoramic view and (b) view from intrados.

The various techniques can often be used in combination, design choices also being influenced by construction phases and requirements regarding possible closure to road traffic. For example, methods requiring work on the extrados may be considered for road bridges but cannot be countenanced for railway bridges, to avoid traffic interruptions.

The most common techniques used for strengthening old masonry barrel vaults are:

- thickening of the old masonry arch with new layers of bricks (Figure 33);
- application of FRP strips (Figures 33 and 34) at the extrados of the barrel vault (Valluzzi, Valdemarca, & Modena, 2001);
- methods of masonry restoration, such as grout injections, repointing of stone joints with good-quality hydraulic lime mortar, crack stitching and patch repairs by manual methods;
- construction of internal brick spandrel walls connected to the extrados of the vault (Figure 35). The new walls are stiff elements which tend to oppose antimetric deformation of vaults, contribute to bearing some of the load and enhance seismic resistance. Lateral spandrel walls have the same effects, since they work as rigid load-bearing walls after rehabilitation and retrofitting of connections with arches (Tecchio et al., 2012).

Some common applications introducing resistant systems are:

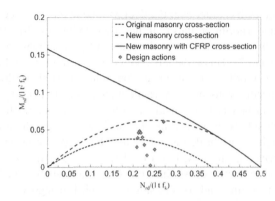

Figure 34. Sandro Gallo Bridge. Safety domain of arch section. M_{rd}, bending moment; N_{rd}, axial load; t, height of section; l, width of section; $f_{M,k}$, compressive strength of masonry.

- saddling (laying of a new RC slab) at the extrados of the vault; RC jacketing at the vault intrados and anchorage with (usually high strength) steel bars and
- prefabricated steel liners at the intrados to support the vaults.

Spandrel walls are generally critical in masonry bridges, because of their high vulnerability to out-of-plane actions. A significant increase in resistance can be obtained by the simple insertion of transversal stainless steel ties (Figure 35), which prevent them from over-turning (Oliveira & Lourenço, 2004). For rehabilitation of piers and abutments, in addition to traditional methods for masonry restoration, masonry post-tensioning techniques

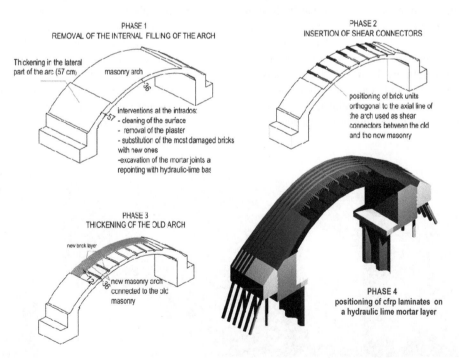

Figure 33. Sandro Gallo Bridge. Thickening of old masonry arch with new layer of bricks and application of FRP strips at extrados of barrel vault.

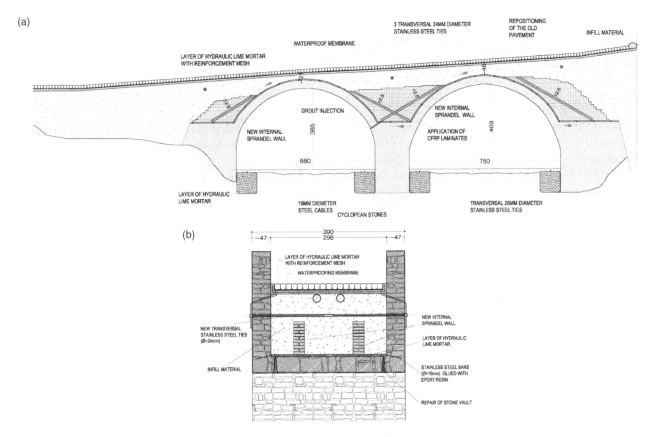

Figure 35.   Rio Moline Bridge. (a) Longitudinal and (b) transversal cross-section.

and jacketing works can be applied. Improved safety levels of foundations, underpinning and new foundations on micro-piles are often used (Figure 33).

These retrofit design solutions and techniques are detailed below, with reference to applications to the masonry bridges studied herein:

Sandro Gallo Bridge, retrofit intervention. Upgrading involved using the existing structure, strengthened with innovative and traditional materials, to enhance current load-bearing capacity for asymmetric load conditions. The intrados of the masonry arch was restored in a 'traditional' way, with cleaning of the surface, removal of plaster, replacement of the most badly damaged bricks, excavation of deteriorated parts of mortar joints, repointing with hydraulic lime-based mortar and repositioning of plaster. The following construction phases were carried out at the extrados (Figure 33):

- removal of the internal filling of the arch;
- preparation of a horizontal level for positioning the concrete foundation beam;
- execution of the new lateral foundation beam on micro-piles (diameter 200 mm, internal tubular reinforcement, hollow section $\phi = 101.6$ mm, 10 mm thick). The new RC edge beam was moulded

in a saw-tooth shape, to help transfer of the internal thrust to the new structures;
- construction near the springing of a new masonry arch layer, regularising the extrados structure and connected to the old masonry;
- thickening of the masonry structure in the central part of the span and positioning of brick units orthogonal to the axial line of the arch, used as connectors between old and new masonry; positioning of 20-mm steel rods, with the same function, attached to the old structure with epoxy resin;
- preparation of the upper surface of the arch and placing of CFRP: removal and replacement of damaged bricks, excavation of deteriorated parts of the mortar joints, repointing with the same hydraulic lime-based mortar used at the intrados, application of a hydraulic lime-based mortar layer, smoothing of the external surface, positioning of CFRP fibres with previous application of primer and epoxy adhesive and final protecting cover;
- refilling of the upper part of the arch with the same material previously removed to reach the road level;
- conclusion of the first working phase, with transfer of the work site to the symmetric part of the bridge, to keep one traffic lane open, alternating one way.

The strengthening effect of retrofitting, with particular reference to asymmetric load configurations, is shown in Figure 34. Checking of the bridge structures included the passage of a conventional truck of 600 kN. A first verification step consisted of analysing only the masonry structures, completed with new brick units, in the central part of the span and at the springing. The second step examined the connection of the CFRP fibres to the bridge structure. For safety evaluation, the load combination presenting a three-point load at one-quarter of the span was applied. Assuming a rectangular stress-block diagram for ultimate stresses on the masonry and elastic behaviour for the CFRP, an ultimate strain of 0.006 and a compressive strength of 2.50 MPa, evaluated according to flat jack test results, were applied to the masonry. M_{rd} and N_{rd} were obtained according to the formulations of Triantafillou (1998), representing the bending moment and axial load defining the safety domain shown in Figure 34.

Rio Moline Bridge, retrofit intervention. Repairs essentially involved rehabilitation of the existing structure by providing specific strategies to recover structural integrity of arches (Figure 35) and to improve element connections, to upgrade seismic resistance. Many of these refurbishment techniques came from the field of historical heritage restoration, the structural capacity of the existing structure being preserved and enhanced with strengthening which was chemically and mechanically compatible with the original ancient structure.

General restoration of the masonry walls (spandrel walls, piers, abutments, wing walls) applied traditional techniques (see Figure 36), such as:

- injection of grout based on hydraulic lime (particularly suitable for consolidation of masonry with cavities);
- repointing of stone joints with hydraulic lime-based mortar and
- local manual reconstruction of masonry.

Repair and strengthening of the stone vaults involved the following procedures:

- temporary removal of pavement and existing fill (with temporary shoring);
- cleaning damaged areas with compressed air and removal of degraded mortar with scrapers;
- positioning of cannulas with a pitch of about 40 cm (at the vault intrados and extrados) and subsequent grout injection;
- applying tensioning wood or plastic wedges;
- repointing of mortar joints;
- construction of internal brick spandrel walls (25 cm thick) connected to the vault extrados and application of CFRP fibres to the lateral surfaces of spandrel walls (after application of primer and epoxy resin);
- insertion of 16-mm stainless steel ties, applied at the vault extrados on a hydraulic lime-based mortar layer and anchored to the lateral spandrel walls. No specific consolidation technique was adopted for foundations, as no evidence of foundation failure was observed and no great increase in loads was expected. Cyclopean stones were used to protect the middle pier from foundation undermining.

After repair (Figure 37), the insertion of internal brick spandrel walls connected to the vault extrados in static conditions contributed to shear some of the load due to concentrated axle forces acting on the vaults. The lateral spandrel walls have the same effect, since they work as

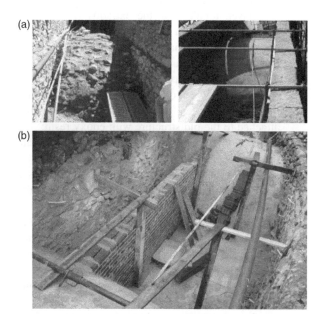

Figure 36. Rio Moline Bridge. (a) Removal of internal fill and consolidation of vaults and (b) construction of internal spandrel walls.

Figure 37. Rio Moline Bridge. Panoramic view with piers, abutments and wing walls restored with traditional techniques.

rigid load-bearing walls after grout injection and restoration of effective connections with arches.

The effectiveness of strengthening can also be evaluated by observation of the improved behaviour under antimetric seismic loads. The Rio Moline Bridge may be viewed as having an infinite rigid spring line, i.e. seismic response in longitudinal direction depends only on the arch, where the possible failure mechanism is localised. Spandrel walls act as rigid elements which can rotate and translate compatibly with the ultimate mechanism characterised by the formation of four hinges in the vault (Figure 38).

The effects of spandrel walls were examined by kinematic analysis. Such walls have a stabilising effect, contrasting antimetric deformation with their in-plane stiffness and leading to a slight increase in triggering acceleration a_0^* and ultimate displacement d_u^* (Figure 38).

Gresal Bridge, retrofit intervention. Work primarily aimed at improving the seismic resistance of the bridge, relying entirely on the intrinsic load-bearing capacity and design characteristics of the existing structure, which was preserved, and even enhanced, in its original configuration. Repairs were carried out in several phases:

- A thin portion of the internal infill layer was removed, to save as much as possible of the fill material with the best mechanical properties, since it has a stabilising function, maintaining the vault voussoirs under compression.
- A new 25-cm-thick RC slab was cast over the whole length of the bridge and anchored to the abutments.
- RC beams resting on micro-piles were emplaced under the pavement outside the existing masonry abutments. The micro-piles were arranged in two

tilted rows, to transfer action to the ground and oppose abutment overturning.

- High-strength bars (26.5 mm diameter) were placed inside the two central piers, in vertical holes drilled from above, for the entire height of the piers down to the foundations. The bars were also anchored at the top to the RC slab, and the combined action of the slab, vertical bars and 'confined' infill created a new resisting strut-tie scheme in the longitudinal direction. In the transverse direction, the vertical reinforcement enhances pier resistance to combined bending-compressive stress states.
- The spandrel walls of the arches were restored in a 'traditional' way. The surface was cleaned, damaged bricks were replaced, deteriorated parts of mortar joints were excavated and re-pointed with hydraulic lime-based mortar, and 6-mm stainless steel bars were inserted in the mortar joints. This phase ended with the emplacement, at the new RC slab level, of transversal 24-mm stainless steel ties, restraining the tops of the walls and avoiding out-of-plane overturning.

The load-bearing capacity of the bridge was increased by retrofitting in both longitudinal and transversal directions. As regards transverse action, the system with tilted micro-piles and an RC transverse slab act as a stabilising portal at the abutment, and the vertical bars in the piers increase their transverse flexural capacity (Figure 39). In the longitudinal direction, the upper RC slab was anchored downward by the vertical steel bars, so that a confinement effect was obtained in the infill. At the ultimate limit state, the non-symmetric kinematic mechanism of the arches tends to be activated by horizontal forces. At that point, the uplift displacement is contrasted by the confined infill (as an ideal inclined strut), thus increasing the overall resistant longitudinal mechanism capacity (Figure 40).

The effectiveness of the strengthening technique is shown by the increase in safety levels after repairs (see capacity curves in Figure 41). In nonlinear static analyses, the masonry and filling material were modelled as a homogeneous, isotropic material according to the Drucker–Prager failure criterion (Pelà, Aprile, & Benedetti, 2008). The following parameters were adopted:

- for masonry:

$$f_c = 1.50\,\text{MPa}, \quad f_t = 0.1f_c, \quad \sin\Phi = \frac{f_c - f_t}{f_c + f_t} = 0.82,$$

$$c = \frac{f_c f_t}{f_c + f_t}, \quad \tan\Phi = 0.47;$$

- for fill material:

$$f_c = 0.10\,\text{MPa}, \quad f_t = \frac{1}{2}f_c, \quad \sin\Phi = 0.33, \quad c = 0.04.$$

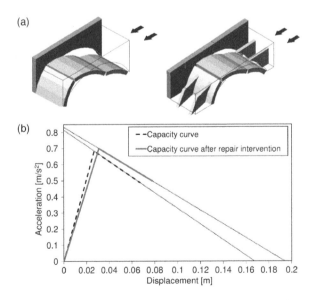

Figure 38. Rio Moline Bridge. (a) Antimetric collapse mechanism before and after insertion of spandrel walls and (b) results of kinematic analysis.

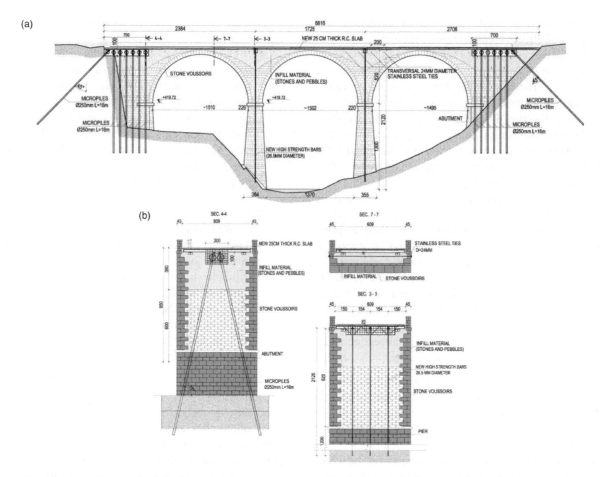

Figure 39. Gresal Bridge. (a) Longitudinal sections with new structural elements (RC slab, micro-piles and vertical ties) and (b) transverse sections on abutments and piers.

The ultimate resistance and ductility of the bridge were increased by retrofitting in both longitudinal and transversal directions. Ductility was the most vulnerable factor in the original structure, mainly due to the slenderness of the high piers. After repairs, as shown in Figure 41, although the elastic stiffness of the bridge appeared to vary less, the ultimate displacement capacity in the inelastic field increased considerably, satisfying in this way the displacement seismic demand.

Lastly, in this case, it should be mentioned that the aforementioned repairs only increased dead loads by about

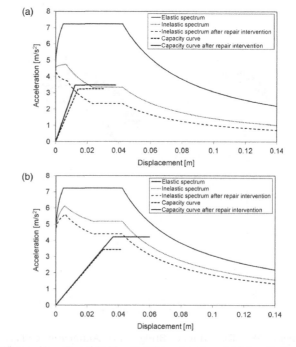

Figure 41. Gresal Bridge. (a) Transversal and (b) longitudinal pushover.

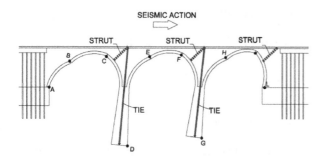

Figure 40. Gresal Bridge. Confining effect on the infill.

1%, and that seismic retrofitting was again related to future functional adaptation, the positioning of the upper RC slab being preliminary to lateral widening of the roadway (anchorages of bars for lateral cantilever structures were installed in the RC slab) and traffic lanes.

4. Conclusions

A considerable percentage of Italian road bridges is of arch type, and masonry bridges are among the oldest structures still in service. Rehabilitation and strengthening interventions for these structures are becoming increasingly necessary due to degradation, worsened by poor maintenance, structural deficiencies with respect to current traffic loads and restrictive safety requirements. In addition, in many seismic regions like Italy, recent updating of national structural codes has made seismic verifications compulsory, with the consequent requirement to plan proper seismic retrofitting of 'strategic' structures for Civil Protection purposes.

In this context, structural engineers are increasingly called upon to devise ways for extending the life of existing bridges while observing severe cost constraints. Establishing principles for assessing and retrofitting common bridge types is essential because, as repair and retrofitting of existing bridges are based on an approach which is substantially different from the design of new structures, knowledge beyond the scope of design codes is required. All this lies at the base of maintenance planning procedures, which at network level should rely on probabilistic life cycle assessment methodologies that optimally allow the prioritisation and allocation of resources for the interventions.

In this paper, the main degradation processes and typical defects of the original designs are presented with reference to the most common types of RC and masonry arch bridges. Rehabilitation of original structures is discussed, and possible techniques for static and seismic retrofitting are illustrated, with reference to significant real cases treated by the authors. In almost all the cases presented, rehabilitation and retrofitting were coupled with refurbishment, to improve safety and comfort to road users. By means of these real case studies, a methodological approach was described with the aim of eventually helping to develop standard intervention procedures on common types of RC and masonry arch bridges included in stocks of Bridge Management Systems.

Notes

1. Email: claudio.modena@unipd.it
2. Email: giovanni.tecchio@dicea.unipd.it
3. Email: marco.dona@dicea.unipd.it
4. Email: francesca.daporto@unipd.it
5. Email: paolo.zampieri@dicea.unipd.it
6. Email: marianoangelo.zanini@dicea.unipd.it

References

BD 79/06, British Standard (2006). *The management of substandard highway structures, bridges and structures—BD series.* Vol. 3, Section 4, Part 18. London: UK Department for Transport.

Carturan, F., Pellegrino, C., Rossi, R., Gastaldi, M., & Modena, C. (2013). An integrated procedure for management of bridge networks in seismic areas. *Bulletin of Earthquake Engineering, 11*, 543–559.

Carturan, F., Zanini, M.A., Pellegrino, C., & Modena, C. (2014). A unified framework for earthquake risk assessment of transportation networks and gross regional product. *Bulletin of Earthquake Engineering, 12*, 795–806.

Clemente, P. (1998). Introduction to dynamics of stone arches. *Earthquake Engineering and Structural Dynamics, 27*, 513–522.

Clemente, P., Occhiuzzi, A., & Raithel, A. (1995). Limit behaviour of stones arch bridges. *ASCE Journal of Structural Engineering, 121*, 1045–1050.

da Porto, F., Franchetti, P., Grendene, M., Valluzzi, M.R., & Modena, C. (2007). Structural capacity of masonry arch bridges to horizontal loads. In P.B. Lourenco, D. Oliveira, & V. Portela (Eds.), *Proceedings of the 5th International Conference on Arch Bridges ARCH'07*, 12–14 September, Madeira.

Gilbert, M., & Melbourne, C. (1994). Rigid-block of masonry structures. *The Structure Engineer, 72*, 356–361.

Heyman, J. (1966). The stone skeleton. *International Journal of solids and Structures, 2*, 249–279.

Heyman, J. (1972). *Coulomb's memoirs on statics.* Cambridge: Cambridge University Press.

ICOMOS (2003). *Recommendations for the analysis, conservation and structural restoration of architectural heritage.* International Scientific Committee for Analysis and Restoration of Structures and Architectural Heritage (ISCARSAH), ICOMOS.

ISO 13822, International Standard (2010). *Bases for design of structures – Assessment of existing structures.* Geneva: International Organization for Standardization.

Italian Ministry of Infrastructures (2008). *Norme Tecniche per le Costruzioni, DM 2008-1-14 (in Italian).* Author.

Modena, C., Franchetti, P., & Grendene, M. (2004). *Guidelines for design of bridges in seismic prone zones.* Mestre, VE: Veneto Strade spa (in Italian).

Oliveira, D.V., & Lourenço, P.B. (2004). Repair of stone masonry arch bridges. In *Proceedings of the 4th International Conference on Arch Bridges ARCH'04*, 17–19 November, Barcelona.

Ordinance of the Presidency of the Council of Ministers No. 3274 (2003). *Primi elementi in materia di criteri generali per la classificazione sismica del territorio nazionale e di normative tecniche per le costruzioni in zona sismica, 2003-03-20 (in Italian).* Rome, Italy: Presidency of the Council of Ministers.

Pelà, L., Aprile, A., & Benedetti, A. (2008). Seismic assessment of masonry arch bridges. *Engineering Structures, 31*, 1777–1778.

Pellegrino, C., da Porto, F., & Modena, C. (2011). Experimental behaviour of reinforced concrete elements repaired with polymer-modified cementitious mortar. *Materials and Structures, 44*, 517–527.

Pellegrino, C., & Modena, C. (2009). Flexural strengthening of real-scale RC and PRC beams with end-anchored pretensioned FRP laminates. *ACI Structural Journal, 106*, 319–328.

Pellegrino, C., & Modena, C. (2010). Analytical model for FRP confinement of concrete columns with and without internal steel reinforcement. *ASCE Journal of Composites for Construction, 14,* 693–705.

Pellegrino, C., & Vasic, M. (2013). Assessment of design procedures for the use of externally bonded FRP composites in shear strengthening of reinforced concrete beams. *Composites Part B: Engineering, 45,* 727–741.

Priestley, M.J.N. (1995). Displacement-based seismic assessment of existing reinforced concrete buildings. In *Proceedings of the Pacific Conference on Earthquake Engineering* (pp. 225–244). Melbourne, Australia: Australian Earthquake Engineering Society.

SB-ICA (2007). Guideline/guidelines for inspection and condition assessment of railway bridges. Prepared by Sustainable Bridges – A project within EU FP6. Retrieved from: http://www.sustainablebridges.net

Tecchio, G., da Porto, F., Zampieri, P., Modena, C., & Bettio, C. (2012). Static and seismic retrofit of masonry arch bridges: case studies. In *Proceedings of 6th International Conference on Bridge Maintenance and Safety (IABMAS'12)*, July 8–12, Stresa, Lake Maggiore, Italy. London: CRC Press, Taylor and Francis.

Triantafillou, T.C. (1998). Strengthening of masonry structures using epoxy-bonded FRP laminates. *ASCE Journal of Composites for Construction, 2,* 96–104.

Valluzzi, M.R., Valdemarca, M., & Modena, C. (2001). Behaviour of brick masonry vaults strengthened by FRP laminates. *ASCE Journal of Composites for Construction, 5,* 163–169.

Zanardo, G., Pellegrino, C., Bobisut, C., & Modena, C. (2004). Performance evaluation of short span reinforced concrete arch bridges. *ASCE Journal of Bridge Engineering, 9,* 424–434.

Zanini, M.A., Pellegrino, C., Morbin, R., & Modena, C. (2013). Seismic vulnerability of bridges in transport networks subjected to environmental deterioration. *Bulletin of Earthquake Engineering, 11,* 561–579.

The art of arches

Man-Chung Tang

People have been building beautiful arch bridges for more than 2000 years. From the spectacularly solid Roman arch to today's elegant and graceful styles, there are a large variety of arch shapes. The development of arches was influenced by the construction materials available at the time. Today, with more versatile materials such as steel and reinforced concrete, many beautiful signature arch bridges are being built.

1. Ancient arches

Arches are probably the oldest form of long-span bridges. The reason for this is simple. Among the four basic categories of bridges in the world (girder bridge, arch bridge, cable-stayed bridge and suspension bridge), the arch is the only bridge form that can be constructed using materials such as stones and bricks that have almost no tensile capacity. In ancient times, stone was the only material available for major construction. Therefore, the arch was the only structural form to employ when a longer span was needed. The form of an arch is very natural. The inspiration behind arch construction may have come from nature. There are an abundance of examples of arch formations in our natural landscape (Figure 1).

Figure 1. Natural arches and a Roman arch bridge.

Today, when we talk about ancient arches, we usually think of Roman arches. No doubt, the Romans were great builders and they built many spectacular arches that were incorporated into palaces, coliseums, bridges and viaducts. However, arch construction can be traced back to before Roman times. One great example of this is the Ishtar Gate of Babylon, built in 600 BC. Even still, at least for bridges, the Romans expanded and perfected the art of arch building. And, after 2000 years, some of their structures are still in relatively good shape. The Romans were master builders and while we do not know how masterful they were at building arch bridges with today's materials, the longest-lasting bridges that we can see today are all stone bridges. And, for obvious reasons, these can only be deck arch bridges, with the bridge deck placed above the arch.

Stone arches are very heavy, the benefit of which is that it gives the arch the capability to resist lateral forces from earthquakes, wind, water and other elements. Because these lateral loads can cause bending moment in the transverse direction, and because bending moment creates both tension and compression stresses in the structure (especially in the piers and foundations), the arch needs this heavy weight to pre-compress the structure, thus overcoming any tensile stress that may arise from these bending moments.

2. Modern arches

Modern arches are much lighter and are mostly built using steel or reinforced concrete. These materials have tensile capacities. A modern arch rib using modern materials can handle compression and tension, as well as bending moment. They are much more slender. The advantages of slenderness and lightness are that they offer excellent opportunities for creating graceful arch forms. They also

31

enable much larger spans. However, the disadvantages of slenderness and lightness are that the arch rib by itself, if not restrained, is usually not sufficiently stable under the required design loads. Because the arch rib is under high compression, it behaves like a beam-column. It can buckle in both in-plane and out-of-plane directions.

Buckling of an arch rib can be treated simply as a stiffness problem. The critical buckling load of a simple column is proportional to the stiffness of its cross section, usually denoted by 'EI', where E is the modulus of elasticity of the material and I is the moment of inertia of the cross section. The buckling of an ach rib is certainly more complicated than that of a simple column, but the basic concept is very similar. It follows that the simplest way to raise the buckling load is to increase the stiffness of the cross section. As the modulus of elasticity is almost constant for any given material, the increase in the stiffness can only be achieved by increasing the moment of inertia, I, of the cross section. The most direct way to increase the moment of inertia of a cross section is to increase the size of the member. One disadvantage of this is that increasing the size of the arch rib would make it appear bulky and, in certain instances, may render the bridge aesthetically undesirable to some.

Figure 2. Fremont Bridge, USA and Guandu Bridge, Taiwan.

There are several ways to stabilise slender arch ribs. In the in-plane direction, the arch ribs are connected to the girder by hangers. Any vertical deformation of the arch rib will force the girder to follow. If the hangers do not elongate, the vertical deformation of the girder must be equal to that of the rib. Thus, the stiffness of both the rib and the girder contribute to the stability of the structure. The axial deformation, that is, the elongation or shortening of the hangers is typically small. Consequently, adding the stiffness of the girder and the stiffness of the arch rib to estimate the buckling load is a good approximation in most cases. As a result, in the case of a stiff girder, the arch ribs can be made more slender. The Fremont Bridge in Portland, Oregon, USA and the Guandu Bridge in Taiwan are good examples (Figure 2).

Another way to increase the in-plane stability of the arch rib is to use inclined hangers. This system is called 'Network Arch Bridge'. In actuality, the network arch bridge acts more like a truss than an arch bridge. The hangers are similar to the diagonals of a truss. Consequently, the arch rib can be made very slender in the in-plane direction (Figure 3).

Figure 3. Network arch bridges. Brandanger Bridge, Norway (courtesy Prof. Per Tveit). and General Macarthur Bridge, Taiwan.

In the out-of-plane direction, the hangers are usually in the same plane as the arch rib, so they cannot provide any assistance to strengthen the arch rib against lateral or out-of-plane buckling. The most common way to solve this problem is to have two arch ribs bracing against each other. The bracing can either be in a crisscross pattern, where the structure will act more like a truss, (Figure 4(a)) or be simply parallel struts, where the structure will act more like a Vierendeel truss (Figure 4(b)). Figure 4(c) shows an interesting way of stabilising the two arch ribs against each other.

(a)

(b)

(c)

Figure 4. Examples of arch rib stabilisation.

(a)

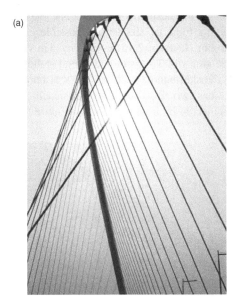

(b)

Figure 5. Spatial arrangement of cables.

Turning the arch and hanger combination into a spatial structure is another effective way of increasing the lateral stability of the arch rib. There are several alternative methods to achieving a similar result: One way is to replace the hangers, which are usually cables, with steel members that have sufficient flexural stiffness in order to restrain the arch rib in the lateral direction (Figure 5(a)). This method can only be efficient in relatively short-span bridges otherwise these beam-hangers would be too long and would become very bulky. Another way is to attach two planes of cables to each arch rib, thus making it into a three-dimensional or spatial structure (Figure 5(b)).

If the bridge is a deck arch, which has no hangers but spandrels, these spandrels underneath the deck can also be inclined, so the deck and the spandrels will form a triangle to stabilise the arch rib. Once the stability problem of an arch rib in both the in-plane and the out-of-plane directions is resolved, the arch can be a very versatile structure for short and medium long-span bridges. In the following section, we will explore a variety of configurations with real-life examples.

3. Caiyuanba Bridge

The Caiyuanba Bridge is located in Chongqing, China. It crosses the Yangtze River from the southern district of the City to the central district. It has a main span of 420 m and is one of the largest arch bridge spans in the world featuring a highway and monorails. The bridge is visible from most areas of the City. Aesthetics was one of the most important criteria employed during design. There are six lanes of city traffic and two pedestrian paths on the upper level of the girder and two monorail tracks on the lower level. The upper deck is 36.50 m wide and the lower deck is 12.10 m wide. This forms a trapezoidal-shaped girder cross section which appears more slender as it crosses the Yangtze River Valley. A rectangular box shape would have looked too bulky at this location.

Because of the aesthetic requirements, the arch ribs were made to look as slender as possible. A simple rectangular box section, 2.40 m wide by 4 m deep, was used for the arch rib. To assure the lateral stability of the arch ribs, a basket-handle arch system was chosen. This arch bridge has two ribs leaning towards each other, and they are connected by transverse struts (Figure 6).

Figure 6.   Caiyuanba Bridge, Chongqing, China.

The water level in the Yangtze River can rise and fall up to 39 m. The entire arch was raised to rest on a pair of piers so that the arch ribs would not be submerged in the water even during the highest high water levels. A partially submerged arch does not look good. As the lower portion of the arch ribs is close to the water level, they are constructed from high-strength concrete to give the bridge better

protection against possible ship collision and corrosion. Aesthetically, the solid lower portion makes the bridge appear more majestic. The basket-handle arch bridge is a good choice for long-span arch bridges. It can also be attractive for short-span bridges if the deck is narrow so that the inclination of the arch ribs is not too acute.

4.   Dagu Bridge

The basket-handle arch configuration is effective, except that with the arches leaning inwards, travellers may feel boxed in while driving through the bridge. This is especially so if the bridge deck is wide and the span is small. To provide them an open view while driving on the bridge, the arch ribs were bent outwards, as in the case of the Dagu Bridge, shown in Figure 7, in Tianjin, China.

Figure 7.   Dagu Bridge, Tianjin, China.

Given this scenario, these arch ribs would have been unstable after the removal of the transverse struts. Consequently, they had to be restrained in the transverse direction by other means. For this reason, each arch rib of the Dagu Bridge has two planes of hangers, forming a spatial structural system. This arrangement provides the required lateral restraint for the arch ribs, which are very slender.

The Dagu Bridge is located in the centre of Tianjin, a city of 11 million close to Beijing. It is one of the first bridges to be built under the Haihe Area Revitalization Plan. Because of its location, aesthetics was extremely important. The owner envisioned a signature structure for the City. The bridge crosses the Haihe River, which traverses the entire city. The river is about 96 m wide at this point; so, the span of the bridge is 106 m to clear the entire river. It carries six lanes of city traffic and two pedestrian/ bicycle paths, one on each side. The minimum width of the bridge deck is 32 m at both ends and varies to about 56 m at the mid-span including some openings in the deck.

Due to navigational clearance requirements, the depth of the girder was limited to about 1.40 m at the centreline of the deck. The 1.40-m girder was not sufficient to span the 32 m width of the bridge in the transverse direction. Hence, the arch ribs were placed inside of the pedestrian/ bicycle paths. They are 24 m apart transverse to the bridge axis. Both arch ribs are made of steel with trapezoidal cross sections. The girder is a steel box girder with an orthotropic steel deck paved with 50 mm of epoxy asphalt.

5. Huihai road Bridge

By moving both arch ribs, like those of the Dagu Bridge, to the centreline of the deck, a new configuration has been derived for the Huihai Road Bridge, in Lianyungang, China (Figure 8). The bridge, currently under construction, is located in the middle of a new development area of the City. Aesthetics for this bridge project is important. The bridge has a main span of 100 m. The deck is 39 m wide. Each of the two arch ribs is restrained by two planes of hangers.

Figure 8. Huihai Road Bridge, Lianyungang, China.

Moving the two arch ribs together changed the behaviour of the structural system. As a comparison, for the Dagu Bridge, the arches are about 24 m apart, and any eccentric load on the deck or any asymmetrical loads acting on the structure can be carried by the two arch ribs to the end supports. There is no need for the girder to be torsionally rigid. However, in the case of the Huihai Road Bridge, when the ends of the two arch ribs come together at the centreline of the main girder, such a system can no longer carry any eccentric load on the deck, or any asymmetrical loads due to wind, earthquakes, etc. Consequently, the main girder must be torsionally stiff to carry all torsional moments induced by any asymmetric loading to the piers. To achieve this, a multi-cell box section is used for the girder to enable the transfer of all torsional moment created by eccentric loads to the supports at the ends of the bridge. The bridge deck is widened in the middle section to offer pedestrians additional space to rest and view the beautiful surroundings.

6. Sanhao Bridge

The configuration for the Sanhao Bridge can be shown by turning the Huihai Road Bridge 90° transversely (Figure 9). The Sanhao Bridge is located in the City of Shenyang. It crosses the Hun River connecting two newly developed communities. The City requested a signature bridge at this location to complement the high-end residential developments on both sides of the river. The bridge has a twin main span of 100 m each and the deck is 34 m wide. The girder depth is limited to 2.40 m to satisfy the bridge's navigational requirements.

Figure 9. Sanhao Bridge, Shenyang, China.

However, budget was limited and the Owner could only afford to build a concrete girder. In China, where costs are concerned, a steel girder is more than twice as expensive as a concrete girder; however, a steel girder is significantly lighter in weight. To employ a concrete girder, a new bridge concept or the 'partially cable-supported girder bridge' was developed (Tang, 2007). This design concept achieves its economic efficiency by fully utilising the carrying capacity of both the cable-stay system and the girder itself. The cables carry only a portion of the weight of the bridge, so the towers can be made much lighter. This design concept also allows a more liberal selection of cable forces, which is a major advantage for such a tower shape. A transverse arch is not a good structure to carry high vertical loads compared to conventional vertical towers. For the design of the Sanhao Bridge, the cables were assigned to carry only about 50% of the load.

The two arch leaves divide each cable into three sections. The force in the middle section of each cable was fine-tuned to eliminate any out-of-plane force in the arch ribs. The two arch ribs were assembled into their final shape by laying them flat on the completed bridge deck and then hoisting them up into their final position using strand jacks attached to a temporary steel frame installed between the two arch ribs.

7. Taijiang Bridge

Yet another new configuration was innovated by combining the two arch ribs, similar to those of the Sanhao Bridge, together into one single arch. This configuration was implemented for the Taijiang Bridge in the City of Sanming in Fujian Province, China (Figure 10).

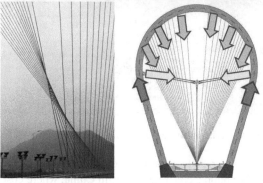

Figure 10.   Taijiang Bridge, Sanming, China.

This bridge has two adjacent main spans of 110 m each. The maximum allowable girder depth is 2.60 m. Again, due to budget constraints, the Owner could only afford to build a concrete girder. The design was based on the concept of a partially cable-supported girder bridge. In this case, as in the Sanhao Bridge, the cables carry approximately 50% of the total load on the bridge. However, putting many cables along the top portion of the arch creates large bending moment in the arch rib. Therefore, a horizontal tie was used to balance the outward thrust, creating a tied arch from the upper part of the arch. This significantly reduces the bending moment in the tower. This horizontal tie consists of two parallel cables of PWS strands. They are very big cables, so there was not sufficient space inside the box-shaped tower ribs to anchor these cables.

Instead, they were anchored to the ribs' inside face of the box with dead-end anchors. To do this, the cables were prefabricated to their exact length after the tower was completely erected, when the exact distance between the anchor points could be determined. A vertical-tie cable was used to tie these horizontal-tie cables to the deck girder. The lower end of this vertical cable was anchored at the bottom of the main girder. The purpose of this vertical-tie cable is to control the force in the horizontal-tie cables. The force in the horizontal-tie cables can be easily estimated after the exact geometry of the horizontal-tie cables is obtained (by surveying) and the force in the vertical-tie cable is calculated. In other words, the force in the horizontal-tie cables can be adjusted by adjusting the force in the vertical-tie cable.

To introduce a little intrigue, the cables are arranged in an inverted position, with the lowest cable anchored to the far end of the bridge. At the lower ends, all cables are anchored along the centreline of the bridge deck. In this way, the arrangement of the cables and the tower are symmetrical to both axes of the bridge. Thus, the tower is always under symmetrical loads, and there is no resultant horizontal force acting on the tower no matter how eccentric the loads on the deck may be. All eccentric or torsional moments are carried by the torsionally rigid box girder. The cables are stressed at the upper end, which is located inside the arch rib.

8. Construction

The arch is a very exciting structural form. The arch is also the most efficient structural system that can carry large loads with minimal construction materials. However, a long-span arch is difficult to build because it is not self-supporting during construction. Most early arch bridges were built using formwork supported from below. This kind of formwork is expensive and the cost increases sharply as the span gets longer (Figure 11).

Figure 11. Construction example of arch bridges: (a) Caiyuanba Bridge; (b) Dagu Bridge; (c) Taijiang Bridge; (d), (e) and (f) raising of arch ribs of Sanhao Bridge.

Most recent long-span arch bridges utilise high lines and cable stays to construct the arch ribs. The high line used for the erection of the Caiyuanba Bridge (Figure 11 (a)) has a lifting capacity of 4.20 MN which is, by far, the largest high line in the world. For medium to short spans, the arch can be constructed using local falsework support, as in the case of the Dagu Bridge (Figure 11(b)). They can also utilise more innovative methods, as in the construction of the Sanhao Bridge (Figure 11(d)–(f)) and the Taijiang Bridge (Figure 11(c)).

9. Conclusion

A few examples of modern arches have been described to show that arch bridges can have a variety of forms. While the arch is the oldest form of long-span bridges, it also represents some of the newest developments in bridge design configurations.

Reference

Tang, M.C. (2007). Rethinking bridge design – A new configuration. *Civil Engineering*, *77*, 38–45.

Innovative steel bridge girders with tubular flanges

Richard Sause

I-shaped steel girders with tubular flanges have been studied for application in highway bridges because of their large torsional stiffness compared to conventional I-shaped steel plate girders (I-girders). For straight girder bridges, the large torsional stiffness of a tubular flange girder (TFG) results in significantly greater lateral–torsional buckling strength compared to a corresponding I-girder. For horizontally curved girder bridges, the large torsional stiffness of a TFG results in much less normal stress, vertical displacement and cross-sectional rotation compared to a corresponding I-girder. The paper presents experimental and finite element analysis results for straight and horizontally curved TFG bridges. The results show the advantages of TFGs in comparison to conventional I-girders. A TFG demonstration bridge constructed in the USA is described.

Introduction

The potential advantages of I-shaped girders with tubular flanges (TFGs) for highway bridges, including large local buckling resistance, large torsional stiffness and reduced web slenderness, were outlined by Wassef, Ritchie, and Kulicki (1997). Wimer and Sause (2004) studied concrete-filled tubular flange girders (CFTFGs) for straight highway bridges. The CFTFGs had a rectangular concrete-filled steel tube as the compression flange and a flat steel plate as the tension flange, as shown in Figure 1(a). The study showed that the large torsional stiffness of the tubular flange enables the use of large unbraced lengths and fewer diaphragms (or cross frames) in the bridge framing system.

Kim and Sause (2005a, 2005b) studied CFTFGs with a round concrete-filled steel tube as the compression flange and a flat steel plate as the tension flange, as shown in Figure 1(b). They summarised several advantages of CFTFGs relative to conventional I-shaped steel plate girders (I-girders) for straight bridges: (1) the concrete-filled tubular flange provides more strength, stiffness and stability than a flat plate flange with same amount of steel, which is a significant advantage under bridge construction conditions before the girders are made composite with the concrete deck (e.g. during construction of the concrete deck) and (2) fewer diaphragms (or cross frames) are needed for CFTFGs to maintain lateral–torsional stability compared to corresponding I-girders, which reduces the effort needed to construct a bridge. Kim and Sause (2005a, 2005b) also provided formulas for determining the flexural

strength of TFGs. Wang, Zhai, Duan, and Li (2009) and Wang, Duan, Zhu, and Wang (2011) conducted tests and analyses of CFTFGs. They observed good composite action between the steel tubular flange and the concrete infill, and they provided design formulas and found good agreement between predicted strengths and test results.

Sause and Dong (2008) studied the application of steel TFGs in curved highway bridges, where the horizontal curvature induces significant torsion to the bridge girder system. The significant torsional stiffness of TFGs, compared to conventional steel I-girders, is the main advantage of TFGs in this application. Sause and Dong (2008) found that, overall, curved TFG systems are more structurally efficient than corresponding curved I-girder systems.

This paper reviews, synthesises and draws conclusions from the previous work on TFG bridges. This work includes tests, finite element (FE) analysis studies, development of flexural strength formulas and construction of a demonstration bridge. Applications in straight and curved highway bridges are considered.

Initial design study

To identify the advantages of TFGs, CFTFGs with minimum steel weight were designed for a straight prototype bridge and compared with minimum weight conventional steel I-girders (Kim & Sause, 2005b). The prototype bridge has a simply supported single span of

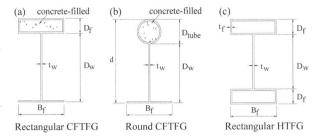

Figure 1. Girders with tubular flanges.

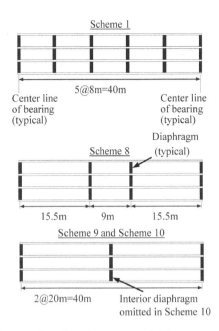

Figure 2. Framing plans for prototype bridge.

40 m and a deck width of 15.2 m. The concrete deck is 254-mm thick with a concrete compressive strength of 27.6 MPa. The bridge has four girders spaced at 3.8 m with 1.9-m deck overhangs. The I-girders and CFTFGs were designed with ASTM A709 HPS 485W steel (with a yield stress of 485 MPa) and were assumed to be composite with the concrete deck in the final constructed condition. The webs have intermediate transverse stiffeners with Category C' fatigue details.

Possible framing plans for the prototype bridge with different diaphragm arrangements are shown in Figure 2. Scheme 1, with diaphragms spaced at 8 m, is typical for a conventional I-girder bridge. Schemes 8 and 9 have fewer interior diaphragms, resulting in reduced fabrication and erection costs. However, the designs of conventional I-girders with such a large diaphragm spacing will be controlled by the compression capacity of the top flange, and in particular by lateral–torsional buckling (LTB) under construction conditions (e.g. during construction of the concrete deck), and will require heavier I-girder cross sections. On the other hand, CFTFGs with their greater top flange strength, stiffness and stability (due to the large torsional stiffness of the tubular flange) efficiently accommodate the larger diaphragm spacing under construction conditions before the girders are made composite with the concrete deck.

The initial design study considered criteria for strength, stability, service and fatigue. For the conventional I-girders, AASHTO LRFD bridge design specifications (American Association of State Highway and Transportation Officials [AASHTO], 1998) were used. For the CFTFGs, AASHTO (1998) provisions for web stability, shear strength and fatigue of I-girders were used without modification. New formulas for TFG flexural strength (presented later) considering LTB and cross-sectional yielding were developed and used.

Figure 3 (Kim & Sause, 2005b) compares the total steel weight of the girders for the prototype bridge, plotted versus the girder depth, for three cases: (1) conventional I-girders with the Scheme 1 diaphragm arrangement; (2) conventional I-girders with the Scheme 8 diaphragm arrangement and (3) CFTFGs with the Scheme 9 diaphragm arrangement. Figure 3 shows that the CFTFGs

are significantly lighter than the I-girders, even with a large diaphragm spacing. In particular, the CFTFGs with Scheme 9 (with only one line of interior diaphragms, as shown in Figure 2) are more than 10% lighter than the I-girders with Scheme 1 (with four lines of interior diaphragms). Thus, the CFTFGs have the advantages of decreased steel weight and decreased diaphragm fabrication and erection effort. As the cost per unit weight to fabricate and erect the diaphragms is significantly greater than the cost per unit weight to fabricate and erect the girders, the decreased diaphragm fabrication and erection effort is expected to lead to significantly decreased total steel fabrication and erection costs for a bridge.

TFG flexural strength

Design flexural strength formulas for TFGs considering LTB and cross-sectional yielding were proposed by

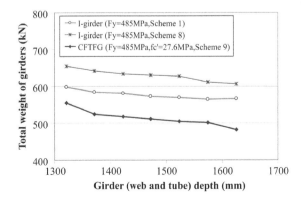

Figure 3. Total steel weight of girders for prototype bridge.

Figure 4. CFTFG test specimen without concrete deck.

Kim and Sause (2005a, 2005b). The formulas were calibrated to fit FE analysis results for CFTFGs (Kim & Sause, 2005a). The resulting design flexural strength M_d is given by:

$$M_d = C_b \alpha_s M_u \leq M_u, \qquad (1)$$

where α_s is the strength reduction factor (see below), M_u is the cross-sectional ultimate flexural strength and C_b is the moment gradient correction factor given by:

$$C_b = \frac{(12.5 M_M)}{(2.5 M_M + 3 M_A + 4 M_B + 3 M_C)}, \qquad (2)$$

where M_M is the absolute value of the maximum moment in the unbraced segment, and M_A, M_B and M_C are the absolute values of the moment at the quarter, middle and three-quarter points of the unbraced length, respectively (American Institute of Steel Construction [AISC], 2005).

The strength reduction factor, accounting for inelastic or elastic LTB, is given by:

$$\alpha_s = 0.8 \left\{ \left[\left(\frac{M_u}{M_{cr}} \right)^2 + 2.2 \right]^{1/2} - \frac{M_u}{M_{cr}} \right\} \leq 1.0, \qquad (3)$$

where M_{cr} is the elastic LTB moment, which is given by:

$$M_{cr} = \frac{\pi E}{L_b / r_y} \sqrt{0.385 K_T A + 2.467 \frac{d^2 A^2}{(L_b / r_y)^2}}, \qquad (4)$$

where E is the modulus of elasticity for steel, L_b is the unbraced length, r_y is the radius of gyration, K_T is the St. Venant torsional constant, A is the transformed cross-sectional area and d is the section depth. The radius of gyration is given by:

$$r_y = \sqrt{\frac{I_{tf} + I_{bf}}{A}}, \qquad (5)$$

where I_{tf} and I_{bf} are the moments of inertia of the top and bottom flanges about the vertical axis, respectively.

The ultimate flexural capacity of the cross section, M_u, is given by:

$$M_u = R_{pc} M_{yc}, \qquad (6)$$

where R_{pc} is the web plastification factor for the compression flange defined in the AASHTO LRFD bridge design specifications (American Association of State Highway and Transportation Officials [AASHTO], 2004). In Equation (6), M_u varies between M_{yc} and M_p. M_p is the cross-sectional plastic moment capacity (accounting for the composite concrete infill in the tube). M_{yc} is the yield moment of the cross section with respect to the compression flange. For CFTFGs, M_{yc} is the smaller of the yield moment calculated using a strain compatibility analysis and the yield moment calculated using a transformed section (TS) analysis to account for the concrete infill (Kim & Sause, 2005a, 2005b).

The flexural strength M_d given by Equation (1) is called the ideal flexural strength (Kim & Sause, 2005a) because it is based on the assumption that each diaphragm provides perfect lateral and torsional bracing at the brace point. Kim and Sause (2005a) show that the effective torsional bracing of TFGs provided by interior diaphragms may not be adequate; adequate torsional bracing would be sufficiently stiff so that LTB could occur only between the brace points. Therefore, Kim and Sause (2005a) recommend calculating a design flexural strength for torsionally braced TFGs M_d^{br} to account for the torsional flexibility of the bracing provided by the interior diaphragms. Formulas for calculating M_d^{br} are given by Kim and Sause (2005a) and are based on the approach described by Yura, Philips, Raju, and Webb (1992).

Tests of TFGs for straight bridges

Tests of CFTFGs (Sause, Kim, & Wimer, 2008) for a straight bridge were conducted considering two different conditions: (1) critical construction conditions, which are the loads and support conditions that occur during construction of the concrete deck, and where the flexural strength of the girders is controlled by the LTB strength and (2) under maximum loads in the final constructed condition, which are the maximum loads and support conditions that occur during the normal use of the bridge, and where the flexural strength of the girders is controlled by the cross-sectional flexural capacity.

The CFTFG test specimen is similar to the 40-m span prototype bridge (Figure 2), but only two girders were included, and the girder dimensions were scaled by 0.45 (Sause et al., 2008). The simply supported test specimen (Figure 4) included a girder with a rectangular tube and a flat web (denoted FWG) and a girder with a rectangular tube and a corrugated web (denoted CWG). The CFTFGs were composite with a concrete deck in the final

Table 1. Test matrix.

Stage	Scheme	Braced by deck	$\dfrac{M_{max}}{M_{const}}$	$\dfrac{M_{max}}{M_{strengthI}}$	$\dfrac{M_{max}}{M_{serviceII}}$	$\dfrac{M_{max}}{M_d}$ FWG	$\dfrac{M_{max}}{M_d}$ CWG
1	9	No	1.00	–	–	0.66	0.56
2	10	No	1.00	–	–	0.98	0.80
3	10	Yes	–	1.00	1.26	0.58	0.73

constructed condition. Intermediate transverse web stiffeners at the quarter points and mid-span, and bearing stiffeners were used in the FWG to control web distortion, based on recommendations by Kim and Sause (2005a). The CWG had bearing stiffeners and one intermediate transverse web stiffener at mid-span.

The test specimen was tested in two different diaphragm configurations: (1) *with two end diaphragms and one interior diaphragm* (Scheme 9) and (2) *with only two end diaphragms* (Scheme 10). Table 1 summarises the test matrix. The Stage 1 test used the Scheme 9 diaphragm arrangement. The Stage 2 and Stage 3 tests used the Scheme 10 diaphragm arrangement.

The Stage 1 and Stage 2 tests simulated the deck construction stage. The deck construction condition was simulated using a distributed load which was gradually increased until the factored design moment was reached at mid-span. Test condition details avoided unintended bracing of the girders by the loading system. Each test began by placing six individual precast concrete deck panels on the girders. Each concrete deck panel was placed on shims on top of the girders without connections between the panels. Teflon pads were inserted between the concrete deck panels and the shims to minimise friction. The crane shown in Figure 5 was used to handle the concrete deck panels and the loading blocks (discussed later) during the tests (Sause et al., 2008).

The Stage 3 test simulated maximum loads in the final constructed condition. The concrete deck was made composite with the CFTFGs as follows. Holes were cored in the precast concrete deck panels, and shear studs were welded to the girders through the holes. The gaps between the concrete deck panels were grouted, and the panels were post-tensioned together longitudinally. Grout was placed

Figure 5. CFTFG test specimen with concrete deck and load.

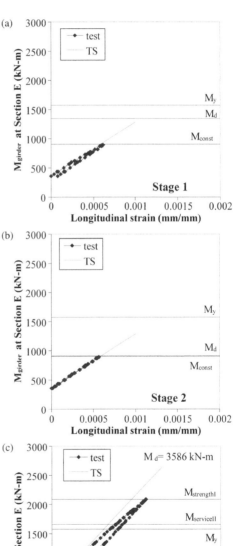

Figure 6. Moment versus longitudinal strain from CFTFG test.

between the concrete deck panels and the girders, and the core holes were filled with grout.

The test loads were based on factored design loads from the AASHTO LRFD bridge design specifications (AASHTO, 1998), and on the design flexural strength formulas presented above. As shown in Table 1, the test loads were quantified by the maximum moment at the mid-span section M_{max} produced by the test loads plus the weight of the girders and the concrete deck. Table 1 compares the estimated M_{max} from the tests with the mid-span moment produced by the factored design loads, as well as the design flexural strength M_d. The subscripts describing the factored design loads are as follows: 'strengthI' represents the loading conditions that produce the maximum load effect on the bridge girders in the final constructed condition; 'serviceII' represents the loading conditions (under normal service conditions) under which permanent deformation of the girders due to steel yielding should be prevented; and 'const' represents the loading conditions that occur in the incomplete bridge during concrete deck construction (Sause et al., 2008).

The design flexural strength M_d for Stage 1 is based on LTB assuming that the interior diaphragm braces the girders torsionally but does not brace the girders laterally, and that the end diaphragms provide perfect lateral and torsional bracing (i.e. M_d equals M_d^{br} described above). M_d for Stage 2 is calculated from Equations (1)–(6) assuming that the end diaphragms provide perfect lateral and torsional bracing and L_b equals the distance between the end diaphragms. M_d for Stage 3 is the cross-sectional flexural capacity M_u because the composite concrete deck provides nearly continuous bracing. In all cases, M_d is based on measured material properties (Sause et al., 2008).

Load was applied to the test specimen by placing 610 × 610 × 1829 mm concrete blocks, with an average weight of 16.0 kN, on the deck. The loading blocks were placed two across on the deck so that the weight of one block could be assumed to act on one girder. The number and spacing of loading blocks was selected to produce a moment distribution similar to that produced by a uniformly distributed load, with a mid-span moment equal to M_{max} in Table 1 (Sause et al., 2008).

Figure 6 compares the test results from the FWG of the test specimen with results from a TS analysis. M_{girder} is the internal resisting moment at a cross section calculated from static equilibrium considering the weights of the loading blocks and the weight of the girders and concrete deck (Wimer & Sause, 2004, Sause et al., 2008). In Figure 6, M_{girder} and the longitudinal normal strain at the bottom flange are taken at a section 152 mm away from mid-span (i.e. 'Section E'). The factored design loads, the design flexural strength M_d and the yield moment M_y are included in Figure 6. M_y is M_{yc}, the yield moment of the CFTFG cross section with respect to the compression flange, which is the smaller of the yield moment calculated

using a strain compatibility analysis and the yield moment calculated using a TS analysis to account for the concrete infill (Kim & Sause, 2005a, 2005b).

The Stage 1 and Stage 2 test data begin after the girders and the concrete deck were placed in the test set-up, so the plots (Figure 6(a),(b)) include an estimate of the initial M_{girder} from the weight of the girders and concrete deck. The Stage 3 test data begin after several loading blocks were placed on the deck to make the stress levels in the test specimen girders the same as in the prototype bridge girders at the time the girders are made composite with the concrete deck. In Figure 6(c), the effect of these loading blocks is included in the estimate of the initial M_{girder}.

The Stage 1 plot (Figure 6(a)) shows some nonlinearity in the loading branch of the test data, due to residual stresses in the steel of the FWG. The flexural rigidity (the slope of the moment-strain plots) from a TS analysis was compared with the test results. For the test data, the slope was calculated from a linear regression line through the unloading branch because of the nonlinearity in the loading branch. The Stage 2 plot (Figure 6(b)) shows that the loading and unloading branches of the Stage 2 test data are essentially identical, because the residual stresses in the steel of the FWG were reduced by loading and unloading during the Stage 1 test. The Stage 3 test results are linear until the stress levels of prior tests are exceeded and are nonlinear thereafter. The flexural rigidity from the test results is about 10% larger than from the TS analysis, and the test results provide validation of the TS analysis approach (Sause et al., 2008).

The CFTFG tests and comparisons with the TS analysis results demonstrate the composite behaviour of the steel TFG cross section with the tube concrete infill. In addition, the Stage 1 and Stage 2 tests demonstrate the significant LTB capacity of TFGs. The tests show that the CFTFGs can carry their design loads. Owing to the loading method employed in the tests, the CFTFGs could not be safely loaded to failure, and the test results could not be used to validate the flexural strength formulas (Equations (1)–(6)). As noted above, however, Kim and Sause (2005a) used FE analysis results to develop and calibrate Equations (1)–(6) for CFTFGs. In addition, Dong and Sause (2009) used FE results to show that Equations (1)–(6) provide an accurate but conservative prediction of the flexural strength of TFGs which do not have tube concrete infill, as discussed below.

Analysis of TFG flexural strength

Dong and Sause (2009) presented FE studies of the flexural strength of TFGs with hollow tubular flanges (HTFGs shown in Figure 1(c)). The tubes and web of these HTFGs are sufficiently compact so that local buckling does not control the flexural strength. The FE models consider material inelasticity, second-order effects, initial

Table 2. HTFG cross sections.

	B_f (mm)	D_f (mm)	t_f (mm)	D_w (mm)	t_w (mm)
CS 20_4	508	101.6	12.7	1066.8	12.7
CS 20_8	508	203.2	12.7	863.6	11.1
CS 20_12	508	304.8	12.7	660.4	7.9

geometric imperfections and residual stresses in the steel tubes of the HTFGs. In the elastic range of the steel, Young's modulus is 200 GPa and Poisson's ratio is 0.3. In the inelastic range, Von Mises yield criteria is used, and a bilinear stress–strain curve is used for the uniaxial stress–strain relationship of the steel. The yield stress of the steel is 345 MPa, and the strain hardening modulus is 2 GPa.

A parametric study performed using the FE models (Dong & Sause, 2009) demonstrated the effects of stiffeners, geometric imperfections, residual stresses, cross-sectional dimensions and bending moment distribution on the LTB flexural strength of HTFGs. These FE results were used to further validate the flexural strength formulas for TFGs (Equations (1)–(6)).

Dong and Sause (2009) studied the effect of cross-sectional distortion on the LTB strength of HTFGs. To study the effect of cross-sectional distortion, a modified FE model was created, which did not permit cross-sectional distortion (denoted as MWOD, model without distortion). In the MWOD, the rotations about the longitudinal axis of all nodes on each cross section are constrained to be equal. The standard model without this constraint is denoted as MWD (model with distortion). The study showed that the use of transverse web stiffeners mitigates the effect of cross-sectional distortion. The results show that three intermediate transverse web stiffeners (at the quarter points and mid-span) and two bearing stiffeners (one at each end of the span), with a thickness of 25 mm, are sufficient to mitigate significant effects of cross-sectional distortion.

This stiffener arrangement was used for the remaining studies. Dong and Sause (2009) also presented parametric studies on the effects of initial geometric imperfections and residual stresses when the flexural strength of an HTFG is controlled by inelastic or elastic LTB. Different initial imperfections were studied. The initial geometric imperfections used in the FE studies to validate Equations (1)–(6) included a maximum top flange initial lateral deflection of $L_b/1000$, where L_b is the unbraced length. $L_b/1000$ was selected for the maximum top flange initial lateral deflection to represent a typical limit on initial out-of-straightness for the compression flange of a bridge girder. The FE studies included a model for residual stresses in the tube of the HTFG, based on the residual stresses identified by Key and Hancock (1993).

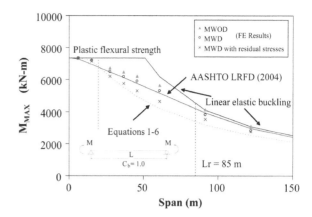

Figure 7. FE results compared with design strength formulas.

To validate Equations (1)–(6), Dong and Sause (2009) studied FE models of HTFGs with three different cross sections shown in Table 2 (see Figure 1(c) for definitions of the dimensions). Figure 7 shows the results from the FE models (Dong & Sause, 2009), plotted versus the slenderness of the HTFG, defined as follows:

$$\lambda = \sqrt{\frac{M_p}{M_{cr}}}, \qquad (7)$$

where M_p is the cross-sectional plastic moment, and M_{cr} is from Equation (4). Figure 7 compares the FE results with the design flexural strength formulas, and it is seen that the formulas from the AASHTO LRFD bridge design specifications (AASHTO, 2004) overestimate the flexural strength from the FE results (MWD with residual stresses), but Equations (1)–(6) (Kim & Sause, 2005a, 2005b) provide an accurate but conservative estimate of the FE results.

Straight TFG demonstration bridge

A two-span straight TFG demonstration bridge spanning Tionesta Creek in Forest County, PA, USA, was designed

Figure 8. TFG demonstration bridge viewed from below.

Figure 9. Pier splice of TFG demonstration bridge.

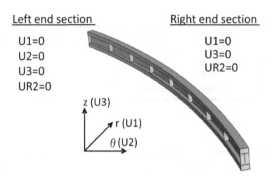

Figure 10. FE model of individual curved HTFG.

and constructed for the Pennsylvania Department of Transportation (Schoedel & Sause, 2010). The TFGs were CFTFGs (Figure 1(c)) with a concrete filled rectangular steel tube as the top flange and a flat steel plate as the bottom flange. The demonstration bridge has two 30.5-m spans and an overall length of 61 m from abutment bearing to abutment bearing. The bridge has two lanes, two shoulders and an overall deck width of 9.55 m. The bridge cross section has four CFTFGs spaced at 2.58 m centres and 0.91 m overhangs (Figure 8). The CFTFGs are made composite with the concrete deck using shear studs. The bridge was designed and constructed as simple spans for dead loads and continuous spans for superimposed dead loads and live loads.

Each span has one line of interior diaphragms near mid-span. End diaphragms are located 0.91 m from the pier centreline in each span. End diaphragms are located at each abutment. The diaphragms are wide flange W21 × 57 beams bolted to connection plates welded to the CFTFGs. The bridge was constructed using ASTM A709 345W weathering steel (with a yield stress of 345 MPa) for the plates and diaphragms, and ASTM A847 weathering steel (with a yield stress of 345 MPa) for the tubes. At the time of the demonstration bridge project, a tube steel with weathering steel characteristics prequalified for welding had not been adopted by AASHTO. Therefore, the project used a tube steel specification requiring the physical properties from ASTM A847, and the chemical composition and energy absorption characteristics from ASTM A588 (Schoedel & Sause, 2010).

The use of only one line of interior diaphragms between the CFTFGs and the use of span-by-span construction enabled rapid erection of the steel girders. The span-by-span construction required a bolted continuity splice at the pier (Figure 9), which was completed

after dead loads of the CFTFGs, diaphragms and concrete deck were carried by the CFTFGs as simple spans. The bolted splice made the girders continuous for superimposed dead loads and live loads (Schoedel & Sause, 2010). The successful completion of the demonstration bridge shows that CFTFGs are a viable steel bridge girder and demonstrates the improved stability of TFGs (compared to conventional I-girders) during steel erection and concrete deck construction. The improved stability enabled several lines of diaphragms to be eliminated that would have been needed for a conventional I-girder bridge.

TFGs for curved bridges

Sause and Dong (2008) studied the application of TFGs to curved highway bridges. Curved steel highway bridges are used widely in the USA. The horizontal curvature induces significant torsion to the bridge girder system, which leads to combined bending and torsion in the girders and complex interaction between diaphragms and girders of the bridge framing system. The significant torsional stiffness of TFGs, compared to conventional I-girders, is the main advantage of TFGs in this application. For individual curved bridge girders, Sause and Dong (2008) found that a TFG develops much less warping normal stress, total normal stress, vertical displacement and cross-sectional rotation than a corresponding curved I-girder. For curved bridge framing systems composed of girders braced by diaphragms between the girders, Sause and Dong (2008) found that fewer (and smaller) diaphragms are needed for the TFG system. Overall, they found that curved TFG systems are more structurally efficient than the corresponding curved I-girder systems.

Dong and Sause (2010b) presented FE studies of individual curved TFGs with hollow tubes (HTFGs), as shown in Figure 1(c). The tubes and web of these HTFGs are sufficiently compact so that local buckling does not influence the strength. A typical FE model is shown in Figure 10, where the 1, 2 and 3 axes are in the radial, circumferential and vertical direction, and U1, U2, U3,

Table 3. Girder cross sections.

Girder	Flanges (mm)	Web (mm)	Depth (mm)	Area (mm^2)
TG1	508 × 101.6 × 12.7	1066.8 × 12.7	1282.7	44,516
TG2	508 × 304.8 × 12.7	660.4 × 7.9	1282.7	46,553
IG1	508 × 25.4	1257.3 × 14.9	1282.7	44,516
IG2	508 × 27.9	1254.8 × 14.5	1282.7	46,553

UR1, UR2 and UR3 are the displacements and rotations about the 1, 2 and 3 axes, respectively. The FE models are simply supported and the girders are braced only at the ends. At each end cross section, the vertical displacements (U3) of the nodes of the bottom wall of the bottom tube are restrained; the lateral (radial) displacements (U1) of the nodes along the line through the web mid-surface are restrained; and the longitudinal (circumferential) rotations about the 2-axis (UR2) of all nodes on the cross section are restrained. The longitudinal (circumferential) displacement (U2) of the centroid of the web is restrained at the left end cross section. The rotations about the 3-axis (UR3) are not restrained at any cross section.

An elastic-perfectly-plastic material is used for the steel of the HTFGs and I-girders. In the elastic range, Young's modulus is 200 GPa and Poisson's ratio is 0.3. In the inelastic range, Von Mises yield criteria is used. The yield stress of the steel is 345 MPa. The FE models are loaded with a vertical load that is uniform across the girder cross section and uniform over the span. Further details of the FE models are given in Dong and Sause (2010b).

In addition to material inelasticity, the FE models (Dong & Sause, 2010b) consider second-order effects, initial geometric imperfections and residual stresses. It was found that initial geometric imperfections and residual stresses had little effect on the load capacity of curved HTFGs. The FE models were used to study the effects of cross-sectional distortion, stiffeners, tube diaphragms and cross-sectional dimensions on the load capacity of curved HTFGs. To study cross-sectional

distortion, a modified FE model (denoted as MWOD) was developed, which did not include cross-sectional distortion. In the MWOD, the rotations about the longitudinal axis (UR2) of all nodes on each cross section are constrained to be equal. The standard model without this constraint is denoted as MWD.

Nonlinear load-displacement analyses of a MWOD and a MWD of a single curved HTFG with a cross section denoted TG2 were conducted (Dong & Sause, 2010b). The dimensions of TG2 are shown in Table 3. The span L (arc length of the girder) is 27 m, the radius of curvature R is 60 m, and the L/R ratio equals 0.45. The load-displacement curves are presented in Figure 11, where the load is normalised by the weight of the girder, and $U3_M$ is the vertical displacement of the mid-span cross section. The results show that the cross-sectional distortion (permitted in the MWD) significantly reduces the load capacity of curved HTFGs (relative to the MWOD), which indicates the importance of this distortion.

Figure 12 shows the deformed cross sections of the MWOD and MWD at mid-span (Dong & Sause, 2010b). The cross-sectional distortion is clearly seen in the MWD results. The tubular flanges are stiff torsionally and the web is comparatively flexible, so the web tends to bend out-of-plane. The tubular flange cross sections also distort.

Further FE studies by Dong and Sause (2010b) show that transverse web stiffeners and tube diaphragms mitigate the effect of cross-sectional distortion shown in Figures 11 and 12. The results show that seven intermediate transverse web stiffeners within the span, a bearing stiffener at each end of the span, and a tube diaphragm within each tube at each end of the span, all

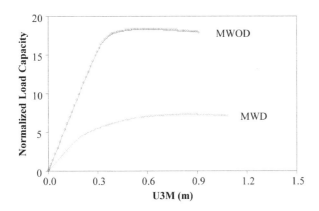

Figure 11. Capacity of individual curved HTFG with (MWD) and without (MWOD) cross-sectional distortion.

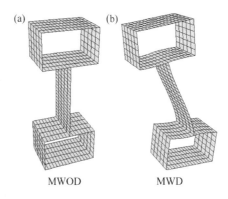

Figure 12. Deformed mid-span cross sections.

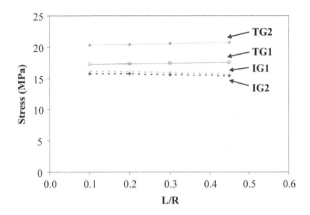

Figure 13.   Maximum bending normal stress.

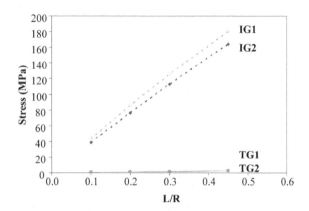

Figure 14.   Maximum warping normal stress.

with thickness of 25 mm, are sufficient to mitigate significant effects of cross-sectional distortion. This stiffener and tube diaphragm arrangement was used for the remaining studies.

The behaviour of individual curved HTFGs was compared with the behaviour of individual curved I-girders. Curved I-girders similar to (corresponding to) the curved HTFGs were developed, denoted IG1 and IG2, with the same girder weight (cross-sectional area), cross-sectional depth and flange width as the corresponding curved HTFGs, denoted TG1 and TG2, as shown in Table 3. The corresponding I-girders have a slightly larger flexural rigidity but much smaller St. Venant torsional rigidity than the corresponding HTFGs (Dong & Sause, 2010b). For the FE studies comparing curved HTFGs with curved I-girders, the span L equals 27 m, and the radius of curvature R varies so that L/R varies from 0.1 to 0.45.

Figure 13 compares the primary bending stress for the curved HTFGs and the corresponding curved I-girders under the girder self-weight (Dong & Sause, 2010b). The primary bending normal stress is calculated from the primary bending moment, which is determined by integrating the normal stresses on the cross section. The primary bending normal stress equals the primary bending moment divided by the section modulus of the cross section about the major axis of the cross section. Figure 13 shows that the I-girders are more efficient than the HTFGs under primary bending moment due to their slightly larger section modulus (about the cross-sectional major axis).

Figure 14 compares the warping normal stress for the curved HTFGs and the corresponding curved I-girders under the girder self-weight (Dong & Sause, 2010b). The warping normal stress is calculated from the lateral bending moment of the top flange, which is determined by integrating the normal stresses on the flange. The warping normal stress is the flange lateral bending moment divided by the section modulus of the flange about the vertical axis. Figure 14 shows that the curved HTFGs have much

smaller warping normal stress than the corresponding curved I-girders, because most of the torsion in the curved HTFG is resisted by St. Venant torsion.

As a result, the flange lateral bending moment from warping torsion of the HTFGs is much smaller than that of the curved I-girders, as shown in Figure 15 (Dong & Sause, 2010b), which shows the variation of the top flange lateral bending moment along the span when $L/R = 0.45$. The curved HTFGs also have a larger flange section modulus (about the vertical axis). Therefore, the maximum warping normal stress for the curved HTFGs is much smaller than that of the curved I-girders (Figure 14).

Figure 16 compares the maximum total normal stress for the curved HTFGs and the corresponding curved I-girders under the girder self-weight (Dong & Sause, 2010b). The results in Figures 13–16 indicate that the bending normal stress is dominant and the warping normal stress is not significant for the curved HTFGs; however, the warping normal stress is dominant for the curved I-girders.

Figures 17 and 18 present the mid-span vertical displacement ($U3_M$) and the mid-span cross-sectional rotation ($UR2_M$) for the curved HTFGs and the

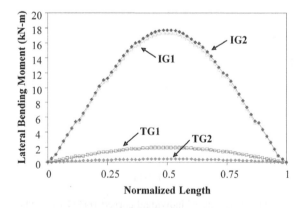

Figure 15.   Top flange lateral bending moment along span.

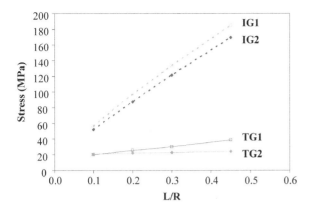

Figure 16. Maximum total longitudinal normal stress.

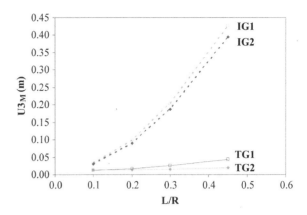

Figure 17. Mid-span vertical displacement.

corresponding curved I-girders under the girder weight (Dong & Sause, 2010b). The curved HTFGs develop much smaller displacement and cross-sectional rotation. For example, for TG2 with $L/R = 0.45$ under girder self-weight, the mid-span vertical displacement $U3_M$ is 0.019 m, which is 1/1500 of the span L, and the cross-sectional rotation at mid-span $UR2_M$ is 0.005 rad. However, for the corresponding curved I-girder IG2 under girder self-weight, $U3_M$ is 0.394 m, which is 1/70 of L, and $UR2_M$ is 0.296 rad. Therefore, the mid-span vertical displacement of the curved I-girder is about 20 times that of the curved HTFG, and the mid-span cross-sectional rotation of the curved I-girder is about 60 times that of the curved HTFG.

In summary, nonlinear load-displacement analyses of individual curved HTFGs show that the effects of initial geometric imperfections and residual stresses on the load capacity of curved HTFGs are small. An arrangement of transverse web stiffeners and tube end diaphragms was identified to mitigate significant effects of cross-sectional distortion on the load capacity. The comparison of individual curved HTFGs with corresponding individual curved I-girders shows that curved I-girders are better at resisting primary bending, but they develop much larger warping normal and total normal stresses, and much larger vertical displacement and cross-sectional rotation than corresponding curved HTFGs, so that, overall, individual curved HTFGs have better structural behaviour than individual curved I-girders.

Under the girder self-weight, the normal stresses, vertical displacement and cross-sectional rotation of the curved I-girders that were studied are quite large, which suggests that temporary support within the span would be needed for these curved I-girders during erection, while the normal stresses, vertical displacement and cross-sectional rotation of the curved HTFGs are quite reasonable, which suggests that such temporary support within the span would not be needed during erection.

Systems of TFGs for curved bridges

Dong and Sause (2010a) studied framing systems for curved highway bridges composed of three HTFGs braced by diaphragms (cross frames). The tubes and web of these HTFGs are sufficiently compact so that local buckling does not control the girder strength. FE models for three-girder systems of curved HTFGs were presented. The FE models were developed considering material inelasticity, second-order effects, initial geometric imperfection and residual stresses. These models were used in parametric studies. The curved HTFG systems were compared with corresponding curved I-girder systems, and the effects of the curvature, cross-sectional dimensions, number of cross frames and a concrete deck were investigated.

Two types of FE models were developed (Dong & Sause, 2010a). The FE model denoted M1 includes the three girders and cross frames, but does not include a composite concrete deck. This model was used to study the behaviour of the three-girder systems during construction of the composite concrete bridge deck. The FE model denoted M2 includes the three-girder system with a composite concrete deck. This model was used to study the behaviour of the three-girder systems in the final constructed condition.

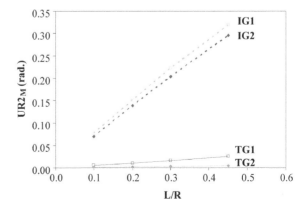

Figure 18. Mid-span cross-sectional rotation.

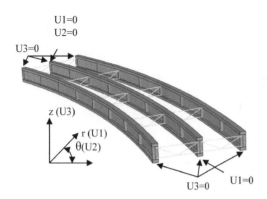

Figure 19. FE model of curved HTFG system.

Table 4. Girder cross-sectional sets.

Set	Girder	Flange (cm)	Web (cm)	Area (cm²)
TGA	G1	40.6 × 20.3 × 1.3	86.4 × 1.0	395.2
	G2	40.6 × 20.3 × 1.3	86.4 × 1.0	395.2
	G3	40.6 × 20.3 × 1.3	86.4 × 1.0	395.2
TGB	G1	40.6 × 20.3 × 1.0	86.4 × 1.0	314.5
	G2	45.7 × 15.2 × 1.3	96.5 × 1.0	407.7
	G3	50.8 × 30.5 × 1.3	66.0 × 0.8	465.3
TGC	G1	30.5 × 15.2 × 1.0	96.5 × 1.1	284.5
	G2	40.6 × 20.3 × 1.0	86.4 × 1.0	320.0
	G3	50.8 × 30.5 × 1.6	66.0 × 1.0	579.0
IGA	G1	40.6 × 2.5	125.7 × 1.5	395.2
	G2	40.6 × 2.5	125.7 × 1.5	395.2
	G3	40.6 × 2.5	125.7 × 1.5	395.2
IGC	G1	40.6 × 1.9	125.7 × 1.3	314.5
	G2	45.7 × 2.5	125.7 × 1.4	407.7
	G3	50.8 × 2.5	125.7 × 1.7	465.3
IGD	G1	30.5 × 1.9	125.7 × 1.3	284.4
	G2	40.6 × 1.9	125.7 × 1.3	319.9
	G3	50.8 × 3.2	125.7 × 2.0	579.0

Model M1 is shown in Figure 19, where the 1, 2 and 3 axes are in the radial, circumferential and vertical direction, and U1, U2, U3, UR1, UR2 and UR3 are the displacements and rotations about the 1, 2 and 3 axes, respectively. The FE model is simply supported. At each end cross section of all three girders, the vertical displacements (U3) of the nodes of the bottom wall of the bottom tube are restrained. At each end cross section of the middle girder G2, the lateral (radial) displacements (U1) of the nodes of the bottom wall of the bottom tube are restrained. The longitudinal (circumferential) displacement (U2) of the centroid of the bottom wall of the bottom tube is restrained at the left end section of G2. To develop Model M2, elements for the concrete deck are added to those shown in Figure 19. The element types used in models M1 and M2 are described by Dong and Sause (2010a).

The material model used for the steel of the HTFGs and I-girders is the same as described above for the studies of individual curved TFGs. The yield stress is 345 MPa. An empirical stress–strain model for unconfined concrete developed by Oh and Sause (2006) is used for the uniaxial stress–strain relationship of the deck concrete. The deck concrete has 31 MPa compressive strength. For both models, M1 and M2, a distributed vertical load is applied uniformly over the span. The magnitude of the uniform load is given as a multiple of a reference load. The reference load is the total weight of the three girders plus the concrete deck. Each girder carries one-third of the total weight of the three girders, and this load is uniform across the girder cross section. Further details of the FE models are given in Dong and Sause (2010a).

The dimensions of the HTFG cross sections in the three-girder systems that were studied are given in Table 4 (Dong & Sause, 2010a), where the cross sections are organised into various 'cross-sectional sets'. Each cross-sectional set includes the cross-sectional dimensions for each girder in the three-girder system, where G1 is the girder on the inside of the curve, G2 is the middle girder and G3 is the girder on the outside of the curve. Cross-sectional sets TGA, TGB and TGC are used for the curved HTFG systems, and cross-sectional sets IGA, IGB and IGC are used for the curved I-girder systems. Each of the I-girders from cross-sectional sets IGA, IGB and IGC has the same girder weight (cross-sectional area), cross-sectional depth and flange width as the corresponding HTFG from cross-sectional sets TGA, TGB and TGC (Table 4).

The span L of the middle girder G2 is equal to 27.4 m. The spacing between girders is 3.05 m. The radius of curvature of each girder R is varied, so the L/R ratio is the same for all three girders. The L/R ratio for the three-girder system varies from 0.1 to 0.45 in the study. Three interior cross frames are used within the span, spaced evenly along the span between the cross frames at the bearings.

Dong and Sause (2010a) studied the effect of cross-sectional distortion on the behaviour of three-girder systems of curved HTFGs. The approach was similar to that used by Dong and Sause (2010b) described above, using a standard FE model with cross-sectional distortion permitted (MWD) and a modified FE model with cross-sectional distortion restrained (MWOD). The results of the study show that that cross-sectional distortion (in the MWD) reduces the load capacity of three-girder systems of curved HTFGs (relative to the MWOD). Additional FE results show that using seven intermediate web transverse stiffeners, and bearing stiffeners and tube diaphragms (within the tube) at the end sections (all with a thickness of 25 mm) is sufficient to mitigate significant effects of cross-sectional distortion on the load capacity of curved HTFG systems. This stiffener and tube diaphragm arrangement was used for the remaining studies of curved HTFG systems.

As noted above, Dong and Sause (2010b) showed that the effects of initial geometric imperfection and residual stresses on the load capacity of individual curved HTFGs

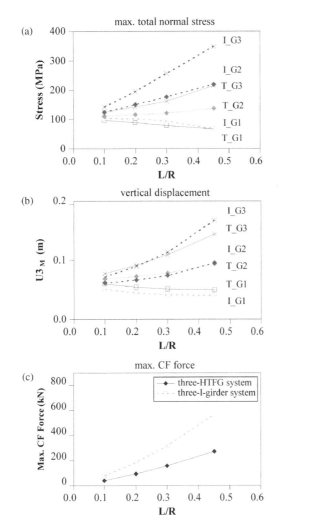

Figure 20.   Results for M1 with section sets TGA and IGA.

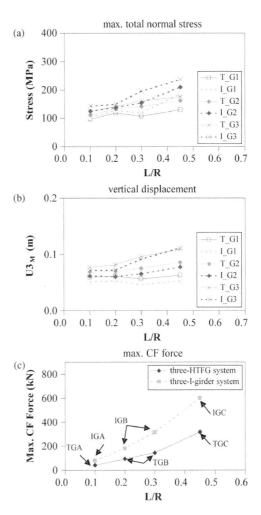

Figure 21.   Results for M1 with section sets varied.

were not significant and could be neglected. Dong and Sause (2010a) conducted a similar study for a system of three curved HTFGs. The results of this study show that the load capacity of the curved HTFG system with initial geometric imperfections and residual stresses is smaller than the load capacity of the system without initial geometric imperfections and residual stresses. However, the differences are small and initial geometric imperfections and residual stresses were not included in the FE models for the remaining studies (Dong & Sause, 2010a).

Using FE analyses, the behaviour of the curved HTFG systems were compared with the behaviour of the curved I-girder system (Dong & Sause, 2010a). Model M1 with cross-sectional set TGA and a similar model M1 with cross-sectional set IGA were analysed under the reference load, which is the weight of the girders and the concrete deck. The maximum total normal stress for each girder in M1 is plotted in Figure 20(a) for the curved HTFGs and the corresponding curved I-girders under the reference load. As the L/R ratio increases, more load is carried by outer girder G3, and less load is carried by inner girder G1. Since the cross sections of

these girders are not varied as the L/R ratio increases, the normal stress increases in girder G3 and decreases in girder G1. For the curved HTFGs, the difference in the maximum normal stress between girders G1–G3 is much smaller than for the corresponding curved I-girders.

Figure 20(a) shows that I-girders G2 and G3 (I_G2 and I_G3) have greater maximum total normal stress than HTFGs G2 and G3 (T_G2 and T_G3). Figure 20(b), however, shows that the vertical displacements of the HTFGs and I-girders are similar, because the HTFGs and I-girders have similar flexural rigidity, and the cross frames and girders work together to resist the torsion acting on the curved bridge system. Figure 20(c) shows that the maximum cross frame force in the curved I-girder system is significantly larger than that in the curved HTFG system.

A second set of FE analyses were performed where larger cross sections were used for G3 and smaller cross sections were used for G1 as the L/R ratio was increased, as follows: Cross-sectional sets TGA and IGA were used when L/R = 0.1, cross-sectional sets TGB and IGB were used when L/R = 0.2 and 0.3, and cross-sectional sets TGC and IGC were used when L/R = 0.45. As shown in

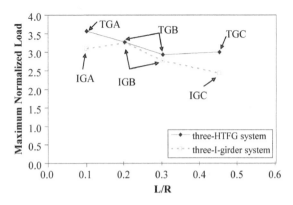

Figure 22. Load capacity of M1 with section sets varied.

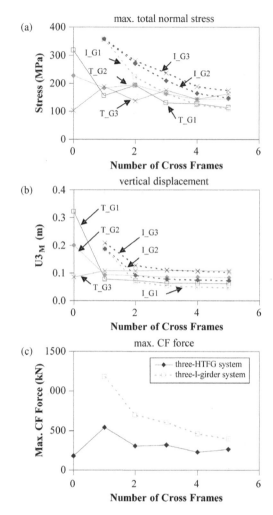

Figure 23. Results for M1 with cross-sectional sets TGC and IGC and $L/R = 0.45$ as number of interior cross frames is varied.

Table 4, each curved I-girder cross section has the same girder weight (cross-sectional area), cross-sectional depth and flange width as the corresponding curved HTFG cross section.

From this second set of FE analyses (Dong & Sause, 2010a) where the cross sections vary with the L/R ratio, the maximum total normal stress, the vertical displacement and the maximum cross frame force under the reference load (i.e. the weight of the girders and the concrete deck) are shown in Figure 21(a)–(c). The normalised load capacity is shown in Figure 22. Similar to the results shown in Figure 20, where cross-sectional sets TGA and IGA are used for all L/R ratios, Figure 21 shows that the HTFG systems develop smaller maximum total normal stress and maximum cross frame force than the I-girder systems. In addition, Figure 22 shows that the HTFG systems have a slightly larger load capacity than the I-girder systems.

When a larger cross section for the outer girder and a smaller cross section for the inner girder are used, the difference in the maximum total normal stress between girders G1–G3 becomes smaller (Dong & Sause, 2010a), as seen by comparing Figures 20(a) and 21(a). However, as the total weight of the girders is constant (i.e. the total weight of the HTFGs with cross-sectional set TGB or TGC equals the total weight of the HTFGs with cross-sectional set TGA), the average maximum total normal stress of the three girders G1–G3 does not change much as the cross-sectional dimensions are varied. These results suggest that if the design of the HTFGs and the corresponding I-girders are optimised for a particular L/R ratio, the maximum normal stress for each of the three girders can be made closer to the average value.

Model M1 with cross-sectional sets TGC and IGC and $L/R = 0.45$ (Dong & Sause, 2010a) was used to study the effect of the number of interior cross frames (equally spaced between the end cross frames at the bearings). The number of interior cross frames is varied from 0 to 5. The maximum total normal stress, the vertical displacement and the maximum cross frame force under the reference load are shown in Figure 23(a)–(c).

The results show that after a single interior cross frame at mid-span is included in the curved HTFG system, the effect on the maximum total normal stress and the vertical displacement is small as more cross frames are added (Dong & Sause, 2010a). After the HTFG system has two interior cross frames, the effect of adding more cross frames on the maximum cross frame force is small. However, the effect of the number of interior cross frames on the curved I-girder system is significant, especially when three or fewer interior cross frames are used. The curved I-girder system without a single interior cross frame cannot carry the weight of the girders and concrete deck. Therefore, Figure 23 does not show results for the I-girder system with no interior cross frames. However, the HTFG system can carry the weight of the girders and concrete deck without interior cross frames.

The above FE analysis results focus on the behaviour of curved girder systems under construction conditions before a composite concrete deck is in place. Dong and Sause (2010a) also studied the behaviour of curved girder systems with a composite concrete deck. FE analyses of

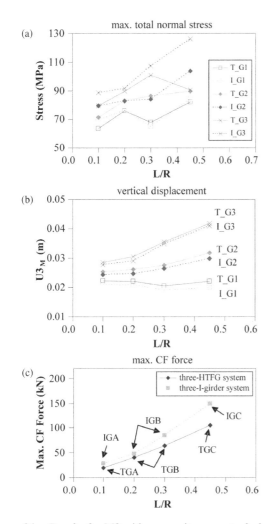

Figure 24. Results for M2 with composite concrete deck.

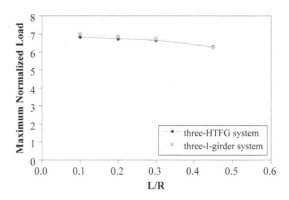

Figure 25.   Load capacity for M2 with composite concrete deck.

shows that with a composite concrete deck in place, the load capacities of the curved HTFG system and the curved I-girder system are similar.

In summary, FE models of three-girder systems of curved HTFGs were developed considering material inelasticity, second-order effects, initial geometric imperfection and residual stresses. The effects of initial geometric imperfections and residual stresses on the load capacity of systems of curved HTFGs were shown to be small. An arrangement of transverse web stiffeners and tube end diaphragms was identified to mitigate significant effects of cross-sectional distortion on the load capacity (Dong & Sause, 2010a).

Comparing the behaviour of curved HTFG systems with corresponding curved I-girder systems demonstrates the following advantages of the HTFG system (Dong & Sause, 2010a): (1) under the same load, the HTFGs develop less total normal stress than the corresponding curved I-girders; (2) the forces in the cross frames of the HTFG systems are smaller than in the corresponding I-girder systems, and thus lighter cross frame members could be used for the HTFG systems; (3) fewer cross frames are needed for the HTFG systems and (4) the HTFG systems can carry their own weight (plus the weight of a concrete deck) without any support within the span and without interior cross frames, and, therefore, temporary support for the HTFG systems during construction (before the concrete deck is composite with the girders) may not be needed, which makes bridge erection faster and less expensive.

Overall, the results indicate that the curved HTFG systems are more structurally efficient than the corresponding curved I-girder systems (Dong & Sause, 2010a). Note that the curved HTFG systems and corresponding curved I-girder systems compared in this paper have the same cross-sectional area (and weight), and, therefore, the material costs for the two systems should be similar. In practice, curved I-girder flanges are often cut from rectangular plates, producing wasted material. On the other hand, curved tubular flanges will be cold-bent from conventional hollow structural shapes, eliminating this wasted material. Fewer cross frames and lighter cross

model M2 of the three-girder system with a composite concrete deck under the reference load (i.e. the weight of the girders and the concrete deck) were conducted. The three-girder systems have three interior cross frames. Cross-sectional sets TGA and IGA are used for $L/R = 0.1$, cross-sectional sets TGB and IGB are used for $L/R = 0.2$ and 0.3, and cross-sectional sets TGC and IGC are used for $L/R = 0.45$. The maximum normal stress, the vertical displacement and the maximum cross frame force from the FE analyses are shown in Figure 24(a)–(c). The normalised load capacity is shown in Figure 25.

For the three-girder systems with the composite concrete deck, the stiffness of the system is significantly increased by the composite deck (Dong & Sause, 2010a). Compared to the results for model M1, the maximum normal stress, the vertical displacement and the maximum cross frame force are reduced, and the load capacity is increased. The curved HTFG system with the composite concrete deck develops smaller maximum total normal stress and maximum cross frame force than the corresponding system of curved I-girders. Figure 25

frame members are needed in HTFG systems. Compared to curved I-girder systems, advantages of curved HTFG systems are expected to be easier handling, shipping and erection of the individual curved girders, due to their significantly increased torsional stiffness (Dong & Sause, 2010b), as well as reduced need for temporary support of HTFG systems during construction, resulting in faster and easier bridge construction.

Summary and conclusions

This paper has reviewed and synthesised previous work on I-shaped steel girders with tubular flanges. TFGs have been studied for application in highway bridges because of their large torsional stiffness compared to conventional I-shaped steel plate girders (I-girders). A design study of CFTFGs for a straight 40-m span simply supported prototype bridge (Kim & Sause, 2005b) was reviewed. The study shows that the CFTFGs are significantly lighter than similar conventional I-girders, even when the diaphragms between the girders are widely spaced. Thus, the CFTFGs have the advantages of decreased steel weight and decreased diaphragm fabrication and erection effort.

Tests of CFTFGs for a straight bridge (Sause et al., 2008) were reviewed. The tests considered: critical construction conditions, during construction of the concrete deck, where the flexural strength is controlled by LTB; and the final constructed condition, where the flexural strength is controlled by the cross-sectional flexural capacity. The tests and comparisons with analysis results demonstrated the composite behaviour of the steel girder cross section and tube concrete infill. In addition, the tests demonstrated the large LTB capacity of TFGs.

Design flexural strength formulas for TFGs considering LTB and cross-sectional yielding were summarised. These formulas were initially calibrated to FE analysis results for CFTFGs by Kim and Sause (2005a). An FE study (Dong & Sause, 2009) of the flexural strength of HTFGs was reviewed. The study considered material inelasticity, second-order effects, initial geometric imperfections and residual stresses. A few intermediate transverse web stiffeners along with bearing stiffeners were found to mitigate significant effects of cross-sectional distortion. Initial geometric imperfections and residual stresses were included in these FE analysis studies that validated the design flexural strength formulas for HTFGs.

A two-span straight CFTFG demonstration bridge, designed and constructed in the USA (Schoedel & Sause, 2010), was described. The TFGs had a concrete filled rectangular steel tube as the top flange and a flat steel plate as the bottom flange. The bridge has two 30.5-m spans. Each span has a single line of interior diaphragms at mid-span. The CFTFGs are made composite with the concrete deck using shear studs. This bridge demonstrated the advantages of the improved stability of TFGs (compared to I-girders) during steel erection and concrete deck construction.

Studies of TFGs in horizontally curved highway bridges (Dong & Sause, 2010a, 2010b) were reviewed. A large torsional stiffness is the main advantage of TFGs in this application, where the curvature induces significant torsion to the bridge girder system. FE studies of individual curved HTFGs (Dong & Sause, 2010b) were reviewed. The FE models considered material inelasticity, second-order effects, initial geometric imperfections and residual stresses. The FE studies show that cross-sectional distortion can reduce the load capacity of an individual curved HTFG, but using intermediate web transverse stiffeners in addition to bearing stiffeners and tube diaphragms (within the tube) at the end sections mitigate significant effects of cross-sectional distortion (Dong & Sause, 2010b).

A comparison of individual curved HTFGs with corresponding I-girders shows that individual curved HTFGs have better structural behaviour than the I-girders. Under the girder self-weight, the normal stresses, vertical displacement and cross-sectional rotation of the curved I-girders were quite large, which suggests that temporary support within the span would be needed for these I-girders during erection, while the normal stresses, vertical displacement and cross-sectional rotation of the curved HTFGs were not large, which suggests that such temporary support within the span would not be needed during erection.

An FE study of curved highway bridge girder systems composed of three HTFGs braced by cross frames (Dong & Sause, 2010a) was reviewed. The FE models of the girders were similar to those used for FE studies of individual curved TFGs. The behaviour of curved HTFG systems compared with the behaviour of corresponding curved I-girder systems is as follows: (1) under the same load, the HTFGs develop less total normal stress than the corresponding I-girders; (2) the forces in the cross frames of the HTFG systems are smaller than in the corresponding I-girder systems, and thus lighter cross frames could be used for the HTFG systems; (3) fewer cross frames are needed for the HTFG systems and (4) the HTFG systems can carry their own weight (plus the weight of a concrete deck) without any support within the span and without interior cross frames, and, therefore, temporary support for the HTFG systems during construction may not be required.

In conclusion, the paper summarises advantages of TFGs compared with conventional I-girders. For straight bridges, the tubular flange of a TFG provides more stability than a flat plate flange with same amount of steel, so that fewer diaphragms (or cross frames) are needed to maintain lateral–torsional stability (before girders are braced by the concrete deck) compared to corresponding I-girders. For curved bridges, the large torsional stiffness of individual curved TFGs, compared to corresponding I-girders, is a significant advantage. Individual curved TFGs develop significantly less warping normal stress, total normal stress, vertical displacement and cross-sectional rotation than corresponding I-girders. After bridge steel erection is completed, and the system of curved I-shaped girders braced by diaphragms (or cross frames) is in place, the curved TFGs develop less normal stress and require fewer (and lighter) diaphragms than corresponding I-girders during construction of the concrete deck.

Note that the TFGs and corresponding I-girders compared in this paper have the same cross-sectional weight (or in some cases the TFGs are lighter). Therefore, the material costs for the two girder types should be similar. In practice, curved I-girder flanges are often cut from steel plates, producing wasted plate material. On the other hand, curved tubular flanges can be cold-bent from conventional hollow structural shapes to avoid this wasted material. Fewer diaphragms (or cross frames) and lighter diaphragms (or lighter cross frame members) are required for TFG systems, resulting in reduced fabrication and erection costs. As the cost per unit weight to fabricate and erect the diaphragms is significantly greater than the cost per unit weight to fabricate and erect the girders, the decreased diaphragm fabrication and erection effort can be expected to lead to significantly decreased total steel fabrication and erection costs for a bridge. Other expected advantages of curved TFG systems compared to corresponding I-girder systems are easier handling, shipping and erection of the individual curved girders, owing to their large torsional stiffness, as well as reduced need for temporary support of curved TFG systems during construction, resulting in faster and easier bridge construction.

Acknowledgement

Michael Baker Corporation of Coraopolis, PA, USA, engineered the superstructure of the TFG demonstration bridge. The contributions of B.G. Kim, M. Wimer, R. Schoedel, and J. Dong to the work reviewed in the paper are gratefully acknowledged. Opinions and conclusions given in the paper are the author's and do not necessarily reflect the views of those acknowledged herein.

Funding

Financial support for this work provided by the Federal Highway Administration (FHWA), the Pennsylvania Department of Transportation (PennDOT) and the Pennsylvania Department of Community and Economic Development through the Pennsylvania Infrastructure Technology Alliance (PITA) is gratefully acknowledged.

References

American Association of State Highway and Transportation Officials (1998). *AASHTO LRFD bridge design specifications*. Washington, DC: Author.

American Association of State Highway and Transportation Officials (2004). *AASHTO LRFD bridge design specifications*. Washington, DC: Author.

American Institute of Steel Construction (2005). *Specification for steel buildings*. Chicago, IL: Author.

Dong, J., & Sause, R. (2009). Flexural strength of tubular flange girders. *Journal of Constructional Steel Research, 65*, 622–630.

Dong, J., & Sause, R. (2010a). Behavior of hollow tubular flange girder systems for curved bridges. *Journal of Structural Engineering, 136*, 174–182.

Dong, J., & Sause, R. (2010b). Finite element analysis of curved tubular flange girders. *Engineering Structures, 32*, 319–327.

Key, P.W., & Hancock, G.J. (1993). A theoretical investigation of the column behaviour of cold-formed square hollow section. *Thin-Walled Structures, 17*, 31–64.

Kim, B.G., & Sause, R. (2005a). *High performance steel girders with tubular flanges* (ATLSS Report No. 05-15) Bethlehem, PA: ATLSS Engineering Research Center, Lehigh University.

Kim, B.G., & Sause, R. (2005b). High performance steel girders with tubular flanges. *International Journal of Steel Structures, 5*, 253–263.

Oh, B., & Sause, R. (2006). Empirical models for confined concrete under uniaxial loading. *International symposium on confined concrete, SP-238*, American Concrete Institute, Farmington Hills, MI, 141–156.

Sause, R., & Dong, J. (2008). Finite-element analysis of curved hollow tubular flange girders. *Proceedings of 25th annual international bridge conference*, Pittsburgh, PA, USA.

Sause, R., Kim, B.G., & Wimer, M.R. (2008). Experimental study of tubular flange girders. *Journal of Structural Engineering, 134*, 384–392.

Schoedel, R.M., & Sause, R. (2010). Design of concrete filled tubular flange girders. *Proceedings of 27th annual international bridge conference*, Pittsburgh, PA, USA.

Wang, C., Duan, L., Zhu, J., & Wang, X. (2011). Innovative research of high performance and sustainable steel bridges. *Proceedings of international symposium on innovation & sustainability of structures in civil engineering*, Xiamen University, Xiamen, Fujian Province, China.

Wang, C.-S., Zhai, X.-L., Duan, L., & Li, B.-R. (2009). Flexural limit load capacity test and analysis for steel and concrete composite beams with tubular up-flanges. *Proceedings of 12th international symposium on tubular structures*, Shanghai, China.

Wassef, W.G., Ritchie, P.A., & Kulicki, J.M. (1997). Girders with corrugated webs and tubular flanges – An innovative bridge system. *Proceedings of 14th annual international bridge conference*, Pittsburgh, PA, USA.

Wimer, M., & Sause, R. (2004). *Rectangular tubular flange girders with corrugated and flat webs* (ATLSS Report No. 04-18) Bethlehem, PA: ATLSS Engineering Research Center, Lehigh University.

Yura, J.A., Philips, B., Raju, S., & Webb, S. (1992). *Bracing of steel beams in bridges* (Research Report No. 1239-4F) Austin, TX: Center for Transportation Research, University of Texas.

Effects of post-failure material behaviour on redundancy factor for design of structural components in nondeterministic systems

Benjin Zhu and Dan M. Frangopol

This paper investigates the effects of post-failure material behaviour on redundancy factor for the design of structural components in nondeterministic systems. The procedure for evaluating the redundancy factors of components of ductile and brittle nondeterministic systems is demonstrated using systems consisting of two to four components. The effects of the number of brittle components in a mixed (ductile–brittle) system and the post-failure behaviour factor on the redundancy factor are also studied using these systems. An efficient approach for simplifying the system model in the redundancy factor analysis of brittle systems with a large number of components is proposed. In order to generate standard tables to facilitate the design process, the redundancy factors and the associated component reliability indices of nondeterministic ductile and brittle systems with large number of components are calculated considering three correlation cases and two probability distribution types. Finally, a bridge example is used to demonstrate the application of the redundancy factor in the design of steel girders taking into account their post-failure behaviour.

Notations

A, B, D, F:	event
$E_c(R)$:	mean resistance of a single component
$E_{cs}(R)$:	mean resistance of a component in a system
g:	performance function
G:	performance function matrix
M:	bending moment
N:	number of components in a system
N_s:	number of series components in a system
N_p:	number of parallel components in a sub-parallel system
N:	normal distribution
LN:	lognormal distribution
P:	load
R:	resistance of a component
V:	coefficient of variation of a random variable
β_c:	reliability index of a single component
β_{cs}:	reliability index of a component in a system
β_{sys}:	reliability index of a system
η_R:	redundancy factor
ϕ_s:	system factor
ϕ_R:	system factor modifier
ρ:	correlation coefficient
δ:	post-failure behaviour factor

1. Introduction

Structural systems are designed to maintain adequate levels of serviceability and safety. In general, redundancy is introduced to some extent in the design of structures and infrastructures to ensure the required serviceability and safety levels. Research on structural redundancy has been extensively performed in the recent decades (Cavaco, Casas, & Neves, 2013; Frangopol & Curley, 1987; Frangopol & Klisinski, 1989a, 1989b; Frangopol & Nakib, 1991; Fu & Frangopol, 1990; Ghosn & Moses, 1998; Ghosn, Moses, & Frangopol, 2010; Liu, Ghosn, Moses, & Neuenhoffer, 2001; Rabi, Karamchandani, & Cornell, 1989; Tsopelas & Husain, 2004; Wen & Song, 2004). Several redundancy measures were proposed to quantify the ability of a structural system to redistribute the applied loads after the failure of its critical members. Frangopol and Curley (1987) defined the system redundancy as the ratio of the reliability index of the intact system, β_{intact}, to the difference between β_{intact} and the reliability index of the damaged system, β_{damaged}. In Rabi et al. (1989), the redundancy is defined by comparing the probability of system collapse to the probability of member failure. Ghosn and Moses (1998) proposed three reliability measures of redundancy that are defined in terms of relative reliability indices. Because redundancy is regarded as an important performance indicator, studies on the evaluation of system redundancy in real structures have been conducted. Tsopelas and Husain (2004) studied the system redundancy in reinforced concrete buildings. Wen and Song (2004) investigated the redundancy of special moment resisting frames and dual systems under seismic excitations. Kim (2010) evaluated the redundancy of steel box-girder bridge by using a nonlinear finite element model.

Although redundancy is a desired structural property, it is not common to find guidance in structural design codes on how to incorporate redundancy in the design process quantitatively. With better understanding of structural behaviour and system reliability-based design (Ang & Tang, 1984; Thoft-Christensen & Baker, 1982; Thoft-Christensen & Murotsu, 1986), Load and Resistance Factor Design (LRFD) replaced Allowable Stress Design within the last decades. LRFD approach (Babu & Singh, 2011; Ellingwood, Galambos, MacGregor, & Cornell, 1980; Hsiao, Yu, & Galambos, 1990; Lin, Yu, & Galambos, 1992; Paikowsky, 2004) provides an improved rational basis to incorporate system reliability and redundancy concepts as it considers the uncertainties associated with the resistance and load effects using separate factors. American Association of State Highway and Transportation Officials LRFD Bridge Design Specifications (AASHTO, 1994, 2012) are examples of a design code that incorporates redundancy in the design process quantitatively.

These specifications account for redundancy by means of a factor (η_R) from the load effects side in the strength limit state equation for a bridge component. Based on a rough classification of component redundancy level, generic values (0.95, 1.0 and 1.05) were recommended for this redundancy factor (AASHTO, 1994, 2012). Nevertheless, selecting the appropriate value for this factor is a very complex task. Furthermore, this redundancy factor does not account for several parameters including system modelling type (i.e. series, parallel), correlation among the resistances of components and post-failure behaviour of components (Ghosn & Moses, 1998; Hendawi & Frangopol, 1994). These parameters must be considered in establishing redundancy factors for design.

Zhu and Frangopol (2014b) investigated the reliability of systems considering the post-failure behaviour (i.e., ductile, brittle) of their components. In addition, Zhu and Frangopol (2014a) provided redundancy factors for systems of various types (i.e. series, parallel) associated with different correlation cases. Redundancy factor, as a multiplier of the resistance rather than the load effects, was defined by Zhu and Frangopol (2014a) as the ratio of the mean resistance of a component in a system when the system reliability is prescribed to the mean resistance of the same component when its reliability index is the same as that of the system. However, the post-failure behaviour of components was not considered by Zhu and Frangopol (2014a) in the evaluation of this redundancy factor. This is the main topic of this paper.

As indicated previously, Zhu and Frangopol (2014b) investigated the effects of post-failure material behaviour on system reliability without considering its effects on system redundancy. The post-failure behaviour of components should be considered in establishing the system redundancy factors for several reasons. First of all, failure of bridge systems is associated with inelastic deformations and the ability of components carrying loads beyond their elastic limits should be considered in evaluating redundancy factors. Second, systems that consist of ductile components are more redundant than those with brittle components. In reality, structural systems may assemble a combination of ductile and brittle components. The material behaviour of each component must be incorporated in redundancy evaluation of the overall system. Finally, the availability of alternative load paths within a system depends on the ductility capacity of its components. Identifying the redistribution of the loads in a system with excessively loaded components and the new load paths after partial failure requires the information on component post-failure behaviour. Therefore, it is essential to consider the post-failure behaviour of components in determining redundancy factors.

This paper investigates the redundancy factor of different systems considering the post-failure behaviour of components. Nondeterministic systems consisting of two to four components are used to illustrate the procedure for calculating the redundancy factors of ductile and brittle systems and to study the effects of the number of brittle components in a mixed (ductile-brittle) system and the post-failure behaviour factor on the redundancy factor. Most structures in practical cases consist of dozens or hundreds of components. The redundancy factors and the associated component reliability indices of nondeterministic ductile and brittle systems with large number of components are calculated with respect to three correlation cases and two probability distribution types for the purpose of generating standard tables to facilitate the design process. Finally, a bridge example is investigated to demonstrate the application of the obtained ductile redundancy factors in the design of steel girders.

It should be noted that although the redundancy of parallel systems considering post-failure behaviour have been studied in Hendawi and Frangopol (1994) and Okasha and Frangopol (2010), the maximum number of components of the ductile or brittle systems analysed was four while the parallel systems investigated in this paper consist of a large number of components (i.e. up to 500). Because the complexity of the redundancy analysis increases with the number of components (especially for the brittle parallel system), an approach for estimating the redundancy factor of the brittle parallel system with large number of components is proposed. In addition, the redundancy factor of mixed systems consisting of both ductile and brittle components that were not investigated in previous studies are analysed in this paper.

2. Brief review of the proposed redundancy factor

Before starting the evaluation of the redundancy factor for ductile and brittle systems, its definition and the procedure for calculation are briefly reviewed in this section. A redundancy factor that accounts for the redundancy from the resistance side in the strength limit state was proposed in Zhu and Frangopol (2014a). It was defined as the ratio of the mean resistance of a component in a system when the system reliability is prescribed to the mean resistance of the same component when its reliability index is the same as that of the system; the effects of several parameters on the redundancy factor were studied and conclusions were drawn from the sensitivity analysis.

The procedure for calculating the redundancy factor is summarised as follows:

(1) Given the probability distribution type of resistance R and load P, the mean value of load $E(P)$, the coefficients of variation of resistance and load $V(R)$ and $V(P)$ and the predefined component reliability index β_c, determine the mean resistance of the component $E_c(R)$;

(2) Given the correlations among the resistances of components and the prescribed system reliability index β_{sys}, evaluate the mean resistance of the components in the system $E_{cs}(R)$;

(3) Calculate the redundancy factor $\eta_R = E_{cs}(R)/E_c(R)$.

The redundancy factor η_R proposed in Zhu and Frangopol (2014a), the system factor modifier ϕ_R proposed by Hendawi and Frangopol (1994), the system factor ϕ_s proposed by Ghosn and Moses (1998) and the factor relating to redundancy in strength limit state of the AASHTO bridge design specifications are of the same nature because all of them serve as a reward or penalty factor. Bridges with sufficient redundancy are rewarded by allowing their components to have less conservative designs while bridges that are non-redundant are required to have higher component capacities.

However, the definitions of these factors are different. The system factor in Ghosn and Moses (1998) is calculated by comparing the strength reserve ratios associated with different limit states with the required redundancy ratios that are determined through a system reliability calibration to provide adequate redundancy. The system factors are calibrated for girder beam bridges with up to 12 beams and box beam bridges with up to 11 box girders. Three different values based on a general classification of the redundancy levels are recommended for the factor relating to redundancy in the AASHTO specifications (AASHTO, 1994, 2012). As mentioned previously, the redundancy factor proposed in Zhu and Frangopol (2014a) is defined as the ratio of the mean resistance of a component when the system reliability index is prescribed to the mean resistance of the component when its reliability index is

equal to the prescribed system reliability (e.g. 3.5). Compared with other factors, the redundancy factor proposed in Zhu and Frangopol (2014a) is more rational and specific because it takes into account the effects of several parameters, such as system type, correlation among the resistances of components, number of components in the system (up to 500) and probability distribution type of loads and resistances.

3. Redundancy factors of systems considering post-failure material behaviour

The systems investigated in Zhu and Frangopol (2014a) do not consider the post-failure behaviour of components. The limit state equations of components in the systems are similar to that of a single component. The failure modes of systems accounting for the post-failure material behaviour of their components are affected not only by the system type but also by the failure sequence of components. Therefore, the step for identifying the failure modes and the associated limit state equations is more complicated than that associated with the systems without considering the post-failure behaviour. Several systems consisting of two, three and four components are used herein as examples to illustrate the process of evaluating the redundancy factors of systems considering post-failure behaviour.

3.1. Redundancy factors of ductile systems

As mentioned previously, the first step in determining the redundancy factor in a system is to find the mean resistance of its component when the component reliability is prescribed as $\beta_c = 3.5$. Consider a single ductile component whose resistance R and load P are modelled as normally distributed random variables. The coefficients of variation of resistance and load $V(R)$ and $V(P)$, and the mean value of load $E(P)$ are assumed to be 0.05, 0.3 and 10, respectively. Therefore, the mean resistance of component $E_c(R)$ is found to be 21.132.

For a system consisting of two ductile components which are identical with the single component just mentioned, two different systems can be formed: series and parallel. Because failure of the series system can be caused by failure of any component, the redundancy factor of series system is not affected by the post-failure behaviour of the components. Therefore, the evaluation of redundancy factors in this paper is mainly focused on the parallel and series-parallel systems.

For a two-component parallel system subject to load $2P$, the resistances of its ductile components are denoted as R_1 and R_2, respectively. Three correlation cases among the resistances of components are considered herein: (a) $\rho(R_1, R_2) = 0$, no correlation; (b) $\rho(R_1, R_2) = 0.5$, partial

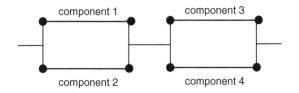

Figure 1. Four-component series-parallel system.

correlation and (c) $\rho(R_1, R_2) = 1.0$, perfect correlation. The statistical parameters associated with R and P are the same as those associated with the single component mentioned previously. By using the Monte Carlo Simulation (MCS) based method described in Zhu and Frangopol (2014a), the mean resistances $E_{cs}(R)$ of ductile components in the two-component parallel system associated with the three correlation cases are found to be 20.810, 20.950 and 21.132, respectively. The corresponding redundancy factors η_R are 0.985, 0.991 and 1.0, respectively. Consequently, the reliability indices of components in the system β_{cs} associated with three correlation cases are 3.40, 3.45 and 3.50, respectively.

Next, the three- and four-component parallel system and the 2×2 series-parallel system (Figure 1) are studied. The loads applied on these systems are $3P$, $4P$ and $2P$, respectively. By performing the same procedure used in the two-parallel system, the mean resistances, redundancy factors and reliability indices of components associated with three- and four-component parallel and series-parallel systems are presented in Table 1.

It is noticed from the results associated with two- to four-component systems that (a) as the number of components in the parallel system increases, the redundancy factor and component reliability index decrease slightly in the no correlation and partial correlation cases; (b) increasing the correlation among the resistances of components leads to higher redundancy factor and component reliability index in the parallel system and (c) compared with the four-component parallel system, the redundancy factor and component reliability index associated with the series-parallel system are higher in the no correlation and partial correlation cases.

3.2. Redundancy factors of brittle systems

Contrary to ductile components, brittle components do not take loads after their fracture failure; therefore, the applied loads will distribute to other remaining components in brittle systems. Due to this property, different failure sequences lead to different load distributions and thus different failure modes in brittle systems. All the possible failure modes must be accounted for and the associated limit state equations need to be identified to determine the redundancy factors. The two-component parallel system described in the previous section is used herein to demonstrate the procedure for calculating the redundancy factors of brittle systems.

Assuming both components are brittle, two different failure modes are anticipated and their respective limit state equations are given as follows:

$$g_1 = R_1 - P = 0, \quad g_3 = R_2 - 2P = 0 \quad (1)$$

and

$$g_2 = R_2 - P = 0, \quad g_4 = R_1 - 2P = 0. \quad (2)$$

Assuming the same statistical parameters of the normally distributed resistances and load as those described previously (e.g. $V(R_i) = 0.05$; $V(P) = 0.3$), the mean resistances of brittle components in the two-component parallel system associated with the three correlation cases are 21.585 if $\rho(R_i, R_j) = 0$; 21.481 if $\rho(R_i, R_j) = 0.5$ and 21.132 if $\rho(R_i, R_j) = 1.0$. The associated redundancy factors are 1.021, 1.017 and 1.0, respectively.

Similarly, the failure modes of three-component parallel systems can be identified, as shown in Figure 2. The limit state equations associated with all the failure modes are:

$$g_1 = R_1 - P = 0, \quad g_2 = R_2 - P = 0,$$
$$g_3 = R_3 - P = 0, \quad (3)$$

$$g_4 = R_2 - 1.5P = 0, \quad g_5 = R_3 - 1.5P = 0,$$
$$g_6 = R_1 - 1.5P = 0, \quad (4)$$

Table 1. $E_{cs}(R)$, η_R and β_{cs} of three- and four-component ductile systems associated with normal distribution.

	System		
Correlation	Three-component parallel system $E_{cs}(R)$; η_R; β_{cs}	Four-component parallel system $E_{cs}(R)$; η_R; β_{cs}	Four-component 2×2 series-parallel system $E_{cs}(R)$; η_R; β_{cs}
$\rho(R_i, R_j) = 0$	20.699; 0.980; 3.37	20.660; 0.978; 3.36	21.160; 1.001; 3.51
$\rho(R_i, R_j) = 0.5$	20.910; 0.989; 3.44	20.893; 0.989; 3.43	21.231; 1.005; 3.53
$\rho(R_i, R_j) = 1$	21.132; 1.000; 3.50	21.132; 1.000; 3.50	21.132; 1.000; 3.50

Note: $V(R) = 0.05$; $V(P) = 0.3$; $\beta_c = 3.5$; $\beta_{sys} = 3.5$; $E_c(R) = 21.132$.

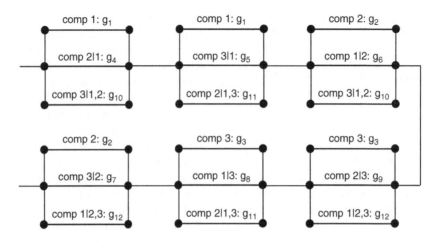

Figure 2.   Failure modes of three-component brittle parallel system.

$$g_7 = R_3 - 1.5P = 0, \quad g_8 = R_1 - 1.5P = 0,$$
$$g_9 = R_2 - 1.5P = 0, \tag{5}$$

$$g_{10} = R_3 - 3P = 0, \quad g_{11} = R_2 - 3P = 0,$$
$$g_{12} = R_1 - 3P = 0. \tag{6}$$

The redundancy factors of the three-component parallel system associated with three correlation cases when R and P follow normal distribution are found to be 1.033 if $\rho(R_i, R_j) = 0$; 1.026 if $\rho(R_i, R_j) = 0.5$ and 1.0 if $\rho(R_i, R_j) = 1.0$.

It is observed from the results related to the two- and three-component brittle systems that (a) the redundancy factor of the parallel system becomes smaller as the correlation among the resistances of components increases and (b) in the no correlation and partial correlation cases, the redundancy factors associated with the two-component parallel system are less than those associated with the three-component parallel system.

3.3.   Redundancy factors of mixed systems

The systems investigated previously consist of only ductile or brittle components. However, there are some cases where both types of material behaviour are included in the system. One of the examples is the steel truss railway bridge in Kama

River of Russia. Its superstructure consists of multi-span steel trusses while its substructure has many single column piers that are made of stones. Therefore, it is necessary to study the redundancy factors of systems having both ductile and brittle components (called 'mixed systems'). Mixed systems consisting of two, three and four components are used herein to investigate the redundancy factors.

For the two-component mixed parallel system, there is only one combination possible: one component is ductile and the other one is brittle (denoted as '1 ductile & 1 brittle'). As more components are included in the mixed system, the number of combinations increases. For the three-component parallel system, two mixed systems are considered: 1 ductile & 2 brittle and 2 ductile & 1 brittle. Similarly, three mixed systems can be formed for four-component parallel system: 1 ductile & 3 brittle, 2 ductile & 2 brittle and 3 ductile & 1 brittle. For the four-component 2×2 series-parallel system, there are two combinations associated with the 2 ductile & 2 brittle case: (a) 2 ductile & 2 brittle Case A, where two ductile components are located in the same sub-parallel system and (b) 2 ductile & 2 brittle Case B, where two ductile components are located in two sub-parallel systems, as shown in Figure 3. Therefore, four different mixed systems can be formed for the 2×2 series-parallel system: 1 ductile & 3 brittle, 2 ductile & 2 brittle Case A, 2 ductile & 2 brittle Case B and 3 ductile & 1 brittle.

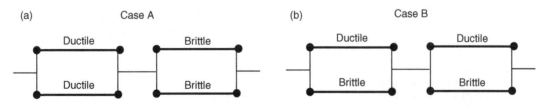

Figure 3.   Four-component series-parallel systems: (a) 2 ductile & 2 brittle Case A and (b) 2 ductile & 2 brittle Case B.

Table 2. $E_{cs}(R)$, η_R and β_{cs} of mixed systems associated with the case $\rho(R_i, R_j) = 0$ when R and P are normal distributed.

System		$E_{cs}(R)$	η_R	β_{cs}
2-Component parallel system	1 ductile & 1 brittle	21.280	1.007	3.55
3-Component parallel system	1 ductile & 2 brittle	21.630	1.024	3.65
	2 ductile & 1 brittle	21.300	1.008	3.55
4-Component parallel system	1 ductile & 3 brittle	21.850	1.034	3.71
	2 ductile & 2 brittle	21.640	1.024	3.65
	3 ductile & 1 brittle	21.319	1.009	3.56
4-Component series-parallel system (2 × 2 SP system)	1 ductile & 3 brittle	21.850	1.034	3.71
	2 ductile & 2 brittle Case A	21.680	1.026	3.66
	2 ductile & 2 brittle Case B	21.680	1.026	3.66
	3 ductile & 1 brittle	21.440	1.015	3.59

Note: $V(P) = 0.3$; $V(R) = 0.05$; $\beta_c = 3.5$; $\beta_{sys} = 3.5$; $E_c(R) = 21.132$.

Assuming the resistances of components and the loads are normally distributed random variables with the coefficients of variation 0.05 and 0.3, respectively, the mean resistances, redundancy factors and reliability indices of components of the mixed systems associated with the no correlation and partial correlation cases are presented in Tables 2 and 3, respectively. In the perfect correlation case ($\rho(R_i, R_j) = 1.0$), $\eta_R = 1.0$ and $\beta_{cs} = 3.5$ for all the mixed systems. It is found from these tables that (a) the redundancy factors of the parallel systems are all at least 1.0 due to the existence of brittle component(s) in the systems and (b) for the 2 × 2 series-parallel system, the redundancy factors associated with the two cases in which the number of brittle components is two are the same; this means that the redundancy factor is not affected by the location of the brittle components in this series-parallel system.

Figure 4 shows the effects of the number of brittle components in the parallel system on the redundancy factor. It is noticed that (a) as the number of brittle components in the parallel system increases, the redundancy factor becomes larger in the no correlation and partial correlation cases and (b) as the correlation among the resistances of components increases, the

redundancy factor increases in the ductile case but decreases in the mixed and brittle cases.

3.4. Effects of post-failure behaviour factor on the redundancy factor

The post-failure behaviour factor δ of a material describes the percentage of remaining strength after failure. The value of δ varies from 0 (i.e. brittle) to 1 (i.e. ductile). The previous sections focus on the redundancy factors associated with only the two extreme post-failure behaviour cases. However, in addition to the ductile and brittle materials, there are some materials whose post-failure behaviour factors are between 0 and 1. Therefore, it is necessary to study the redundancy factors associated with these intermediate post-failure behaviour cases. In this section, parallel systems consisting of two to four components are used to investigate the effects of post-failure behaviour factor on the redundancy factor.

The post-failure behaviour factors of all components are assumed to be the same. The resistances and load associated with the components are considered as normally distributed variables with the coefficients of variation equal to 0.05 and 0.3, respectively. After

Table 3. $E_{cs}(R)$, η_R and β_{cs} of mixed systems associated with the case $\rho(R_i, R_j) = 0.5$ when R and P are normal distributed.

System		$E_{cs}(R)$	η_R	β_{cs}
2-component parallel system	1 ductile & 1 brittle	21.260	1.006	3.53
3-component parallel system	1 ductile & 2 brittle	21.530	1.019	3.62
	2 ductile & 1 brittle	21.290	1.007	3.55
4-component parallel system	1 ductile & 3 brittle	21.700	1.027	3.67
	2 ductile & 2 brittle	21.550	1.020	3.62
	3 ductile & 1 brittle	21.318	1.009	3.55
4-component series-parallel system (2 × 2 SP system)	1 ductile & 3 brittle	21.700	1.027	3.67
	2 ductile & 2 brittle Case A	21.585	1.021	3.63
	2 ductile & 2 brittle Case B	21.585	1.021	3.63
	3 ductile & 1 brittle	21.420	1.014	3.59

Note: $V(P) = 0.3$; $V(R) = 0.05$; $\beta_c = 3.5$; $\beta_{sys} = 3.5$; $E_c(R) = 21.132$.

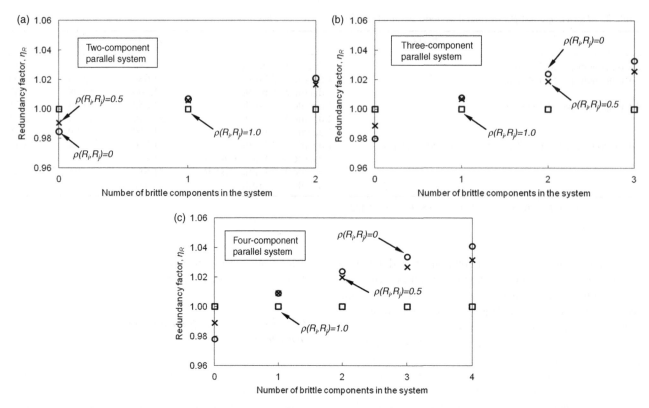

Figure 4. Effects of number of brittle components on the redundancy factor in the parallel systems consisting of (a) two components, (b) three components and (c) four components.

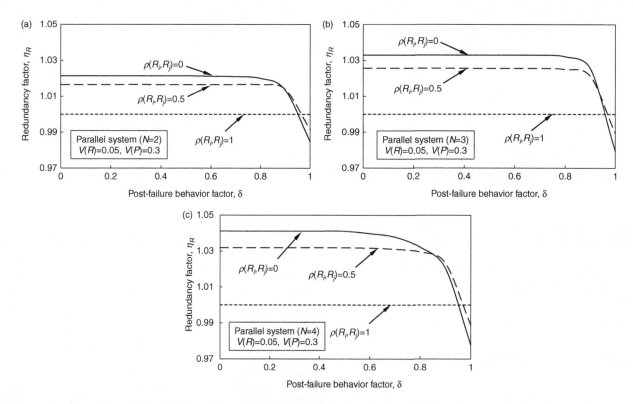

Figure 5. Effects of post-failure behaviour factor δ on redundancy factor η_R in the parallel systems consisting of (a) two components, (b) three components and (c) four components.

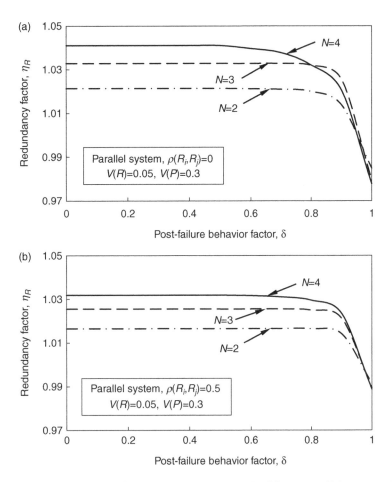

Figure 6. Effects of post-failure behaviour factor δ on redundancy factor η_R in different parallel systems: (a) no correlation case and (b) partial correlation case.

identifying the failure modes of the parallel system and formulating the associated limit state equations, the redundancy factors of the two-, three-, and four-component parallel systems associated with different post-failure behaviour factors are calculated using the MCS-based method. The results are plotted in Figures 5 and 6.

It is noted that: (a) as δ increases from 0 to 1 in the no correlation and partial correlation cases, η_R in the three systems first remains the same and then decreases dramatically; (b) as the correlation among the resistances of components becomes stronger, the region of δ during which η_R remains the same increases; (c) η_R is not affected by δ in the perfect correlation case; (d) the differences in the redundancy factors associated with the three systems are almost the same for $\delta < 0.6$ and become less significant with increasing δ above 0.6 and (e) the redundancy factors reach almost the same value when δ is close to 1.0 (i.e. ductile).

During the calculation of the redundancy factor, the mean resistance of components ($E_{cs}(R)$) when the system reliability index is 3.5 is obtained. Substituting $E_{cs}(R)$ into the component reliability analysis yields the reliability

indices of components. Figures 7 and 8 show the effects of the post-failure behaviour factor on the component reliability index in the three parallel systems associated with three correlation cases. Most of the conclusions drawn from these two figures are similar to those regarding redundancy factors obtained from Figures 5 and 6. Moreover, it is seen that the reliability index of components when $\delta = 0$ (i.e. brittle) is greater than 3.5, while its value when $\delta = 1.0$ (i.e. ductile) is less than 3.5. This is because brittle systems are much less redundant than ductile systems and, therefore, a larger redundancy factor ($\eta_R > 1.0$) needs to be applied to penalise the brittle components by designing them conservatively ($\beta_{cs} > 3.5$); while in the ductile case, smaller redundancy factors ($\eta_R < 1.0$) can be used to achieve a more economical component design ($\beta_{cs} < 3.5$).

4. Redundancy factors of ductile and brittle systems with many components

Redundancy factors associated with the ductile and brittle systems consisting of no more than four components were investigated previously. The results show that η_R is

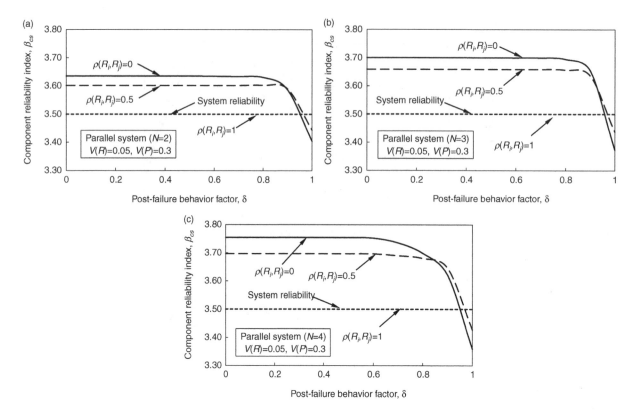

Figure 7.   Effects of post-failure behaviour factor δ on component reliability index in the parallel systems consisting of (a) two components, (b) three components and (c) four components.

affected by the number of components in the system. Although the conclusions obtained from the previous sections provide information about the effects of the number of components on the redundancy factors of ductile and brittle systems with small number of components, it is difficult to determine the redundancy factors for systems with a large number of components based on this information. In this context, it is necessary to evaluate the redundancy factors of ductile and brittle systems that consist of many components so that standard tables of redundancy factors can be generated to facilitate the component design process.

4.1.   Redundancy factors of ductile systems with many components

As stated previously, only the parallel and series-parallel systems are studied in this paper because the redundancy factors of series systems are independent of the material behaviour of its components. The redundancy factors associated with N-component series systems have been provided in Zhu and Frangopol (2014a). Based on the conclusion that the redundancy factor associated with a certain system is not affected by the mean value of the applied load, which was obtained in Zhu and Frangopol (2014a), the following assumption is made for the loads acting on the investigated parallel and series-

parallel systems: (a) for an N-component parallel system, the load it is subject to is $N \cdot P$, where P is the load applied to a single component which is used to calculate $E_c(R)$ and (b) for an N-component series-parallel system that has N_s series component and N_p parallel component ($N_p = 5$, 10 and 20 herein), the load acting on it is $N_p \cdot P$. In this way, the load effect of each component in the intact parallel and series-parallel system is P so that the obtained mean resistance $E_{cs}(R)$ can be compared with $E_c(R)$ to calculate the redundancy factor η_R.

The reliability of ductile systems with many components has been addressed in Zhu and Frangopol (2014b). A ductile component continues to carry its share of the load equal to its capacity after it fails. Therefore, for an N-component ductile parallel system, the load acting on an intact component j after m components in the system fail is $(N \cdot P - \sum_{i=1}^{m} R_i)/(N - m)$. This value is not affected by the failure sequence of the m components. Because the failure modes of ductile systems are independent of the failure sequence, the limit state equation of an N-component parallel system can be written as (Zhu and Frangopol 2014b):

$$g = \sum_{i=1}^{N} R_i - N \cdot P = 0. \qquad (7)$$

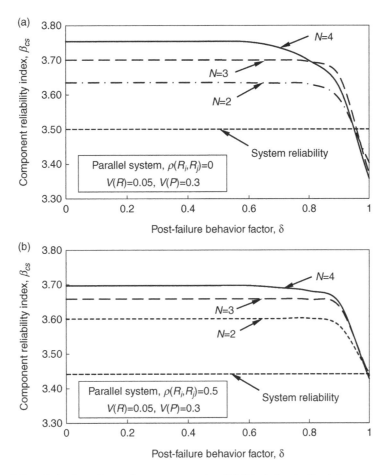

Figure 8. Effects of post-failure behaviour factor δ on component reliability index in the (a) no correlation case and (b) partial correlation case.

For an $N_\mathrm{p} \times N_\mathrm{s}$ series-parallel system which has N_s possible failure modes, the limit state equation associated with failure mode k is (Zhu and Frangopol 2014b)

$$g = \sum_{i=N_\mathrm{p}\cdot(k-1)+1}^{N_\mathrm{p}\cdot k} R_i - N_\mathrm{p} \cdot P = 0, \qquad (8)$$

where $N_\mathrm{p} \cdot (k-1) + 1$ and $N_\mathrm{p} \cdot k$ denote the first and last component in the kth sub-parallel system, respectively. With the identified limit state equations of the N-component ($N = 100, 200, 300, 400$ and 500) ductile parallel and series-parallel systems and the assumed coefficients of variation of resistance and load equal to

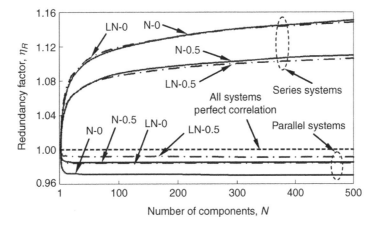

Figure 9. Effects of number of components on the redundancy factor in ductile systems.
Note: 'N' denotes normal distribution; 'LN' denotes lognormal distribution; '0' denotes $\rho(R_i, R_j) = 0$; '0.5' denotes $\rho(R_i, R_j) = 0.5$.

Table 4. $E_{cs}(R)$ and η_R of ductile systems associated with the case $\rho(R_i, R_j) = 0$.

System		Normal distribution		Lognormal distribution	
		$E_{cs}(R)$	η_R	$E_{cs}(R)$	η_R
100-Component system	Series system	23.626	1.118	30.457	1.120
	Parallel system	20.519	0.971	26.759	0.984
	5 × 20 SP system	21.428	1.014	27.928	1.027
	10 × 10 SP system	21.026	0.995	27.439	1.009
	20 × 5 SP system	20.794	0.984	27.085	0.996
200-Component system	Series system	23.921	1.132	30.784	1.132
	Parallel system	20.519	0.971	26.759	0.984
	5 × 40 SP system	21.576	1.021	28.119	1.034
	10 × 20 SP system	21.132	1.000	27.575	1.014
	20 × 10 SP system	20.878	0.988	27.248	1.002
300-Component system	Series system	24.112	1.141	31.028	1.141
	Parallel system	20.498	0.970	26.759	0.984
	5 × 60 SP system	21.639	1.024	28.227	1.038
	10 × 30 SP system	21.195	1.003	27.656	1.017
	20 × 15 SP system	20.921	0.990	27.303	1.004
400-Component system	Series system	24.217	1.146	31.137	1.145
	Parallel system	20.498	0.970	26.759	0.984
	5 × 80 SP system	21.703	1.027	28.255	1.039
	10 × 40 SP system	21.238	1.005	27.711	1.019
	20 × 20 SP system	20.942	0.991	27.303	1.004
500-Component system	Series system	24.323	1.151	31.246	1.149
	Parallel system	20.498	0.970	26.759	0.984
	5 × 100 SP system	21.745	1.029	28.309	1.041
	10 × 50 SP system	21.280	1.007	27.738	1.020
	20 × 25 SP system	20.963	0.992	27.357	1.006

Note: $E(P) = 10$; $V(P) = 0.3$; $V(R) = 0.05$; $\beta_c = 3.5$; $\beta_{sys} = 3.5$; $E_{c,N}(R) = 21.132$; $E_{c,LN}(R) = 27.194$.

0.05 and 0.3, respectively, the redundancy factors associated with two probability distribution types (i.e. normal and lognormal) and three correlation cases (i.e. $\rho(R_i, R_j) = 0$, 0.5 and 1.0) when the system reliability is 3.5 can be obtained using the MCS-based method.

In the perfect correlation case ($\rho(R_i, R_j) = 1.0$), $\eta_R = 1.0$ and $\beta_{cs} = 3.5$ for different types of systems with different number of components associated with both normal and lognormal distributions. This is because for systems whose components are identical and perfectly correlated, the system can be reduced to a single component; therefore, the redundancy factors in this correlation case do not change with the system type and the number of components.

Tables 4 and 5 present the redundancy factors associated with the correlation cases $\rho(R_i, R_j) = 0$ and 0.5, respectively, for the parallel and series-parallel ductile systems along with the results of series systems to facilitate the comparison analysis. In these tables, $E_{c,N}(R) = 21.132$ and $E_{c,LN}(R) = 27.194$ denote the mean resistance of a single component with 3.5 reliability index when its R and P follow normal and lognormal distributions, respectively. These results are also plotted in Figure 9 which shows the effects of number of components on the redundancy factors of series and parallel ductile systems.

It is observed that: (a) the effect of N on η_R in the parallel ductile system depends on the value of N: when N is small ($N \leq 5$), increasing N leads to lower η_R in the parallel system, and the change is less significant as the correlation among the resistances of component increases; however, when $N > 5$, η_R remains almost the same as N increases; (b) for the series-parallel ductile systems that have the same number of parallel components (N_p is the same in these series-parallel systems), η_R increases with N; (c) as the correlation among the resistances of components becomes stronger, η_R decreases and increases in the series and parallel system, respectively and (d) in the series system, the redundancy factors associated with normal and lognormal distributions are very close; however, in the parallel system, the differences in the redundancy factors associated with these two probability distribution cases are more significant.

The component reliability indices β_{cs} of the N-component ductile systems associated with the normal and lognormal cases are shown in Table 6. The results are also plotted in Figure 10 to directly display the effects of N on the component reliability index. It is found that: (a) the effects of N and $\rho(R_i, R_j)$ on the reliability index of components are similar to those on the redundancy factors just discussed and (b) in the series and parallel systems, the component reliability index associated with normal distribution is

Table 5. $E_{cs}(R)$ and η_R of ductile systems associated with the case $\rho(R_i, R_j) = 0.5$.

System		Normal distribution		Lognormal distribution	
		$E_{cs}(R)$	η_R	$E_{cs}(R)$	η_R
100-Component system	Series system	23.013	1.089	29.533	1.086
	Parallel system	20.815	0.985	26.976	0.992
	5 × 20 SP system	21.449	1.015	27.847	1.024
	10 × 10 SP system	21.195	1.003	27.439	1.009
	20 × 5 SP system	21.026	0.995	27.221	1.001
200-Component system	Series system	23.203	1.098	29.777	1.095
	Parallel system	20.815	0.985	26.976	0.992
	5 × 40 SP system	21.555	1.020	27.928	1.027
	10 × 20 SP system	21.259	1.006	27.575	1.014
	20 × 10 SP system	21.090	0.998	27.248	1.002
300-Component system	Series system	23.309	1.103	29.913	1.100
	Parallel system	20.815	0.985	26.949	0.991
	5 × 60 SP system	21.660	1.025	28.010	1.030
	10 × 30 SP system	21.322	1.009	27.602	1.015
	20 × 15 SP system	21.132	1.000	27.330	1.005
400-Component system	Series system	23.414	1.108	29.995	1.103
	Parallel system	20.815	0.985	26.949	0.991
	5 × 80 SP system	21.703	1.027	28.037	1.031
	10 × 40 SP system	21.343	1.010	27.629	1.016
	20 × 20 SP system	21.132	1.000	27.357	1.006
500-Component system	Series system	23.457	1.110	30.077	1.106
	Parallel system	20.815	0.985	26.949	0.991
	5 × 100 SP system	21.703	1.027	28.064	1.032
	10 × 50 SP system	21.364	1.011	27.656	1.017
	20 × 25 SP system	21.132	1.000	27.357	1.006

Note: $E(P) = 10$; $V(P) = 0.3$; $V(R) = 0.05$; $\beta_c = 3.5$; $\beta_{sys} = 3.5$; $E_{c,N}(R) = 21.132$; $E_{c,LN}(R) = 27.194$.

higher and lower than that associated with lognormal distribution, respectively.

4.2. Redundancy factors of brittle systems with many components

As indicated previously, for the evaluation of redundancy factors of brittle systems, all the possible failure modes and associated limit state equations need to be identified and accounted for to perform a correct reliability analysis. An approach to compute the reliability of brittle systems with many components has been formulated in Zhu and Frangopol (2014b). The number of failure modes for an N-component brittle parallel system is N factorial ($N!$). When N is small ($N \le 4$), the approach described in the previous section for determining the failure modes and limit state equations can be used; however, when $N > 4$, the number of failure modes will exceed 120 and it becomes difficult and computationally expensive to consider all the failure modes and associated limit states. Therefore, an alternative approach that can be combined with MATLAB (Mathworks, 2009) is introduced herein.

Table 6. Component reliability index β_{cs} of ductile systems.

System		Normal distribution		Lognormal distribution	
		$\rho(R_i, R_j) = 0$	$\rho(R_i, R_j) = 0.5$	$\rho(R_i, R_j) = 0$	$\rho(R_i, R_j) = 0.5$
100-Component system	Series system	4.23	4.05	3.88	3.77
	Parallel system	3.32	3.40	3.44	3.48
200-Component system	Series system	4.31	4.10	3.91	3.80
	Parallel system	3.32	3.40	3.44	3.48
300-Component system	Series system	4.36	4.14	3.94	3.81
	Parallel system	3.31	3.40	3.44	3.47
400-Component system	Series system	4.40	4.16	3.95	3.82
	Parallel system	3.31	3.40	3.44	3.47
500-Component system	Series system	4.42	4.18	3.96	3.83
	Parallel system	3.31	3.40	3.44	3.47

Note: $E(P) = 10$; $V(P) = 0.3$; $V(R) = 0.05$; $\beta_c = 3.5$; $\beta_{sys} = 3.5$; $E_{c,N}(R) = 21.132$; $E_{c,LN}(R) = 27.194$.

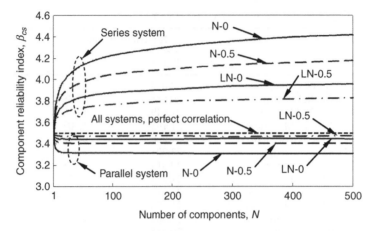

Figure 10. Effects of number of components on the reliability index of components in ductile systems.
Note: 'N' denotes normal distribution; 'LN' denotes lognormal distribution; '0' denotes $\rho(R_i, R_j) = 0$; '0.5' denotes $\rho(R_i, R_j) = 0.5$.

The number of limit state equations of the three-component parallel system is 12, as shown in Equations (3)–(6). It is noticed that some limit state equations can be merged because they are actually the same (i.e. g_4 and g_9, g_5 and g_7 and g_6 and g_8). After merging the identical ones, the limit state equations associated with the three-component parallel system are renumbered as follows:

$$g_1 = R_1 - P = 0, \quad g_2 = R_2 - P = 0,$$
$$g_3 = R_3 - P = 0, \tag{9}$$

$$g_4 = R_1 - 1.5P = 0, \quad g_5 = R_2 - 1.5P = 0,$$
$$g_6 = R_3 - 1.5P = 0, \tag{10}$$

$$g_7 = R_1 - 3P = 0, \quad g_8 = R_2 - 3P = 0,$$
$$g_9 = R_3 - 3P = 0. \tag{11}$$

It is seen that the number of the limit state equations after merging is nine. Similarly, the four-component parallel system has 16 limit state equations after merging.

Therefore, the number of the limit state equations associated with an N-component parallel system is N^2. The failure modes of the three-component parallel system with renumbered limit state equations are shown in Figure 11. It is observed that (a) g_1, g_2 and g_3 correspond to the cases where component 1, 2 and 3 fails first, respectively; (b) g_4, g_5 and g_6 correspond to the cases where component 1, 2 and 3 fails second, respectively and (c) g_7, g_8 and g_9 correspond to the cases where component 1, 2 and 3 fails last, respectively. Therefore, for an N-component brittle parallel system, its limit state equations can be formulated as a matrix (Zhu and Frangopol 2014b):

$$G = \begin{bmatrix} g_1 & g_2 & \cdots & g_N \\ g_{N+1} & g_{N+2} & \cdots & g_{2N} \\ \cdots & \cdots & \cdots & \cdots \\ g_{N(N-1)+1} & g_{N(N-1)+2} & \cdots & g_{N^2} \end{bmatrix}. \tag{12}$$

The element $G(i, j)$ in this matrix denotes that the failure sequence of component j is i. The limit state equation

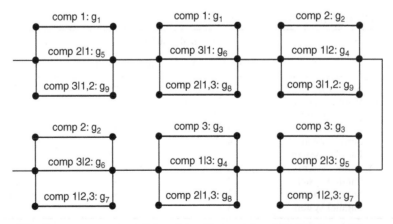

Figure 11. Failure modes of the three-component parallel system with renumbered limit state equations.

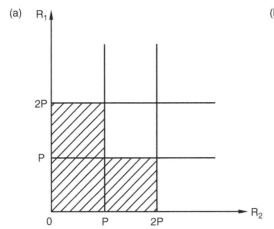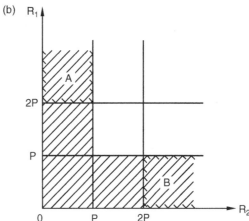

Figure 12.   Sample space of (a) event F_1 and (b) event F_2.

associated with the element $G(i, j)$ in the matrix is given by (Zhu and Frangopol 2014b)

$$G(i,j) = R_j - \frac{N \cdot P}{N - i + 1} = 0, \tag{13}$$

where $i, j = 1, 2, 3, \ldots, N$. For example, the coordinate of the limit state equation g_2 in the matrix is $(1, 2)$ (i.e. first row and second column); therefore, g_2 represents the case where component 2 fails first, and the associated limit state equation is $R_2 - P = 0$ (see Equation (9)). Similarly, $g_{N(N-1)+1}$ stands for the case in which component 1 fails last; when $N = 3$ (three-component system), $g_7 = G(3, 1) = R_1 - 3P = 0$ (see Equation (11)).

After placing all the limit state equations into a matrix, the failure modes of the N-component parallel system can be easily obtained by selecting N elements that are in different rows and columns from the matrix and the set consisting of these selected N elements is one possible failure mode of the system. For example, the limit state equation matrix of the aforementioned three-component parallel system is

$$G = \begin{bmatrix} g_1 & g_2 & g_3 \\ g_4 & g_5 & g_6 \\ g_7 & g_8 & g_9 \end{bmatrix}, \tag{14}$$

where $g_i (i = 1, 2, \ldots, 9)$ are defined in Equations (9)–(11). According to the selection process indicated previously, six possible failure modes can be found from the matrix: (a) $g_1 \rightarrow g_5 \rightarrow g_9$; (b) $g_1 \rightarrow g_6 \rightarrow g_8$; (c) $g_2 \rightarrow g_4 \rightarrow g_9$; (d) $g_2 \rightarrow g_6 \rightarrow g_7$; (e) $g_3 \rightarrow g_4 \rightarrow g_8$ and (f) $g_3 \rightarrow g_5 \rightarrow g_7$. These are the same as the failure modes shown in Figure 11.

By using the limit state equation matrix G, the process of generating limit state equations associated with different failure sequences and identifying the failure

modes can be achieved with MATLAB codes. The procedure for estimating the redundancy factor in brittle systems using this approach is summarised as follows:

(a) Determine the limit state equation of component j when it fails at the sequence i using Equation (13), where $i, j = 1, 2, \ldots, N$;
(b) Form the limit state equation matrix G by defining g_k $(k = 1, 2, \ldots, N^2)$ using the format shown in Equation (12);
(c) Identify all the combinations, each consisting of N elements located in different rows and columns of the matrix G; the obtained combinations are the failure modes of the system;
(d) Based on the obtained limit state equations and failure modes, and other statistical information associated with the resistances and load, the mean resistance of component when the system reliability index is prescribed (i.e. 3.5) can be determined;
(e) Calculate the redundancy factor.

This approach is used to compute the redundancy factor of the brittle parallel systems with up to eight components. However, the nine-component parallel system has 362,880 different failure modes and the reliability analysis becomes very time consuming and in most servers (such as a Dell Precision R5500 rack workstation equipped with two six cores X5675 Intel Xeon processors with 3.06 GHz clock speed and 24 GB DDR3 memory) the memory usage is exceeded. Therefore, in order to calculate the redundancy factors of brittle parallel systems consisting of more than eight components, another method has to be introduced.

Consider the two-component brittle parallel system described previously. It has two different failure modes and the associated limit state equations are shown in Equations (1) and (2). Its system failure can be expressed

Table 7. $E_{cs}(R)$ and η_R of brittle systems associated with the case $\rho(R_i, R_j) = 0$.

System		Normal distribution		Lognormal distribution	
		$E_{cs}(R)$	η_R	$E_{cs}(R)$	η_R
5-Component system	Series system	22.125	1.047	28.581	1.051
	Parallel system	22.125	1.047	28.581	1.051
10-Component system	Series system	22.484	1.064	29.098	1.070
	Parallel system	22.506	1.065	29.125	1.071
	5×2 SP system	22.506	1.065	29.125	1.071
15-Component system	Series system	22.738	1.076	29.342	1.079
	Parallel system	22.738	1.076	29.342	1.079
	5×3 SP system	22.738	1.076	29.342	1.079
20-Component system	Series system	22.865	1.082	29.560	1.087
	Parallel system	22.865	1.082	29.560	1.087
	5×4 SP system	22.865	1.082	29.560	1.087
	10×2 SP system	22.865	1.082	29.560	1.087
25-Component system	Series system	22.992	1.088	29.641	1.090
	Parallel system	22.992	1.088	29.641	1.090
	5×5 SP system	22.992	1.088	29.641	1.090
50-Component system	Series system	23.330	1.104	30.104	1.107
	Parallel system	23.330	1.104	30.104	1.107
	5×10 SP system	23.330	1.104	30.104	1.107
	10×5 SP system	23.330	1.104	30.104	1.107

Note: $E(P) = 10$; $V(P) = 0.3$; $V(R) = 0.05$; $\beta_c = 3.5$; $\beta_{sys} = 3.5$; $E_{c,N}(R) = 21.132$; $E_{c,LN}(R) = 27.194$.

in terms of components failure events as (Zhu and Frangopol 2014b)

$$F_1 = [(g_1 < 0) \cap (g_3 < 0)] \cup [(g_2 < 0) \cap (g_4 < 0)].$$
(15)

Denoting the event $g_i < 0$ as D_i, the above equation can be rewritten as

$$F_1 = (D_1 \cap D_3) \cup (D_2 \cap D_4).$$
(16)

The probability of event F_1 is approximately equal to the probability of the following event F_2:

$$F_2 = (D_1 \cup D_2) \cap (D_3 \cup D_4)$$

$$= [(g_1 < 0) \cup (g_2 < 0)]$$

$$\cap [(g_3 < 0) \cup (g_4 < 0)].$$
(17)

This is explained using Figure 12. The sample spaces generated by the events F_1 and F_2 are shown in Figure 12 (a) and (b), respectively. It is seen that

$$F_2 = F_1 \cup A \cup B,$$
(18)

where event A is $(R_1 > 2P) \cap (R_2 < P)$ and event B is $(R_2 > 2P) \cap (R_1 < P)$, as shown in the Figure 12(b).

Because R_1 and R_2 have the same mean value and standard deviation, the probabilities of occurrence of events A and B are very small and can be neglected. Therefore, the event F_2 in Equation (17) can be used to find the failure

probability of the two-component brittle parallel system. Extending this conclusion to the N-component brittle parallel system yields the system failure event as follows (Zhu and Frangopol 2014b):

$$F = [(g_1 < 0) \cup (g_2 < 0) \cup \ldots$$

$$\cup (g_N < 0)] \cap [(g_{N+1} < 0) \cup (g_{N+2} < 0) \cup \ldots$$

$$\cup (g_{2N} < 0)] \cap \ldots \cap [(g_{N(N-1)+1} < 0)$$

$$\cup (g_{N(N-1)+2} < 0) \cup \ldots \cup (g_{N^2} < 0)].$$
(19)

where $g_1, g_2, \ldots, g_N^2$ are the performance functions listed in Equation (12). Therefore, by simplifying the system model from an $N! \times N$ series-parallel system to an $N \times N$ series-parallel system, the redundancy factors can be computed for brittle parallel systems having a large number of components.

It should be noted that this approach for estimating the failure probability of the brittle parallel system with many components is based on the assumption that the resistances of components in the system are the same. With this assumption, the limit state equations can be merged to form the limit state equation matrix for the failure modes identification and failure probability estimation. In most practical cases, the components in parallel positions are usually designed to have the same (or very similar) dimensions (e.g. pier columns, beam girders). Therefore, if the material of the components is brittle, the system failure probability can be approximately evaluated using this approach.

Table 8. $E_{cs}(R)$ and η_R of brittle systems associated with the case $\rho(R_i, R_j) = 0.5$.

System		Normal distribution		Lognormal distribution	
		$E_{cs}(R)$	η_R	$E_{cs}(R)$	η_R
5-Component system	Series system	21.914	1.037	28.200	1.037
	Parallel system	21.914	1.037	28.200	1.037
10-Component system	Series system	22.189	1.050	28.608	1.052
	Parallel system	22.189	1.050	28.608	1.052
	5×2 SP system	22.189	1.050	28.608	1.052
15-Component system	Series system	22.358	1.058	28.771	1.058
	Parallel system	22.337	1.057	28.771	1.058
	5×3 SP system	22.337	1.057	28.771	1.058
20-Component system	Series system	22.463	1.063	28.880	1.062
	Parallel system	22.442	1.062	28.880	1.062
	5×4 SP system	22.442	1.062	28.880	1.062
	10×2 SP system	22.442	1.062	28.880	1.062
25-Component system	Series system	22.527	1.066	28.962	1.065
	Parallel system	22.527	1.066	28.962	1.065
	5×5 SP system	22.527	1.066	28.962	1.065
50-Component system	Series system	22.759	1.077	29.288	1.077
	Parallel system	22.759	1.077	29.288	1.077
	5×10 SP system	22.759	1.077	29.288	1.077
	10×5 SP system	22.759	1.077	29.288	1.077

Note: $E(P) = 10$; $V(P) = 0.3$; $V(R) = 0.05$; $\beta_c = 3.5$; $\beta_{sys} = 3.5$; $E_{c,N}(R) = 21.132$; $E_{c,LN}(R) = 27.194$.

With the coefficients of variation of resistances and load being 0.05 and 0.3, respectively, the redundancy factors associated with two probability distribution types (i.e. normal and lognormal) and three correlation cases (i.e. $\rho(R_i, R_j) = 0$, 0.5 and 1.0) are calculated with respect to the N-component ($N \le 50$) parallel and series-parallel brittle systems. Similar to the ductile systems, the redundancy factors associated with the perfect correlation case (i.e. $\rho(R_i, R_j) = 1.0$) in the brittle systems are also 1.0. The redundancy factors associated with the other two correlation cases are shown in Tables 7 and 8. Figure 13 plots the effects of number of components on the redundancy factors of brittle series and parallel systems.

It is noted that: (a) the redundancy factors η_R of the brittle parallel systems are all greater than 1.0, which implies that the brittle components have to be designed conservatively ($\beta_{cs} > 3.5$) even in the parallel systems; (b) when the number of brittle components are fixed, η_R associated with series, parallel and series-parallel systems are the same; this indicates that for an N-component brittle structure, η_R is independent of the system type; (c) as the number of components in the brittle system increases, η_R associated with all types of systems become larger; (d) η_R of all types of systems decreases when the correlation among the resistances of components becomes stronger; (e) in the no correlation case, η_R associated with the

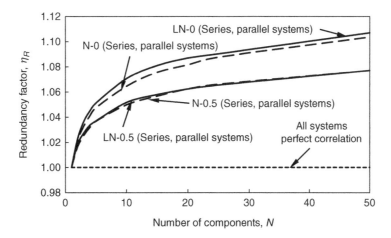

Figure 13.   Effects of number of components on the redundancy factor in brittle systems.
Note: 'N' denotes normal distribution; 'LN' denotes lognormal distribution; '0' denotes $\rho(R_i, R_j) = 0$; '0.5' denotes $\rho(R_i, R_j) = 0.5$.

Table 9. Component reliability index β_{cs} of brittle systems.

System		Normal distribution		Lognormal distribution	
		$\rho(R_i, R_j) = 0$	$\rho(R_i, R_j) = 0.5$	$\rho(R_i, R_j) = 0$	$\rho(R_i, R_j) = 0.5$
5-component system	Series system	3.79	3.73	3.67	3.62
	Parallel system	3.79	3.73	3.67	3.62
10-component system	Series system	3.90	3.81	3.72	3.67
	Parallel system	3.90	3.81	3.72	3.67
15-component system	Series system	3.97	3.86	3.76	3.69
	Parallel system	3.97	3.86	3.76	3.69
20-component system	Series system	4.01	3.89	3.78	3.70
	Parallel system	4.01	3.89	3.78	3.70
25-component system	Series system	4.04	3.91	3.79	3.71
	Parallel system	4.04	3.91	3.79	3.71
50-component system	Series system	4.14	3.98	3.84	3.74
	Parallel system	4.14	3.98	3.84	3.74

Note: $E(P) = 10$; $V(P) = 0.3$; $V(R) = 0.05$; $\beta_c = 3.5$; $\beta_{sys} = 3.5$; $E_{c,N}(R) = 21.132$; $E_{c,LN}(R) = 27.194$.

lognormal distribution case is higher than that associated with the normal distribution case and (f) in the partial correlation case, η_R associated with normal and lognormal distributions are almost the same.

It should also be noted that although the redundancy factors of the N-component series and parallel brittle systems are identical, the designs of components (i.e. the mean resistances) in the two systems are not the same. This is because the loads applied on the series and parallel systems are different when calculating the redundancy factors. The mean resistances of the N-component series and parallel systems listed in the tables are computed with respect to the loads P and $N \cdot P$, respectively. Therefore, when the load is fixed, the mean resistance associated with the brittle parallel system is lower than that associated with the series system, which clearly indicates that the parallel system is more economical than the series system.

The component reliability indices of the N-component brittle systems when the system reliability indices are 3.5

are presented in Table 9 and Figure 14. It is seen that (a) for the brittle systems, increasing the number of components leads to higher reliability indices of components in both series and parallel systems; (b) in the no correlation and partial correlation cases, the reliability indices of components associated with the normal distribution are higher than those associated with the lognormal distribution and (c) as the correlation among the resistances of components increases, the component reliability indices in both series and parallel systems decrease.

5. A bridge example

A bridge example is presented herein to demonstrate the application of the proposed redundancy factor by taking into account the post-failure material behaviour in the design of steel girders. This bridge was used for demonstrating (a) the procedure for evaluating the reliability of systems consisting of equally reliable components (Zhu and Frangopol 2014b) and (b) the

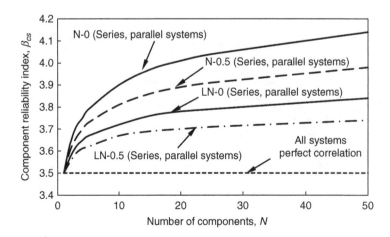

Figure 14. Effects of number of components on the reliability index of components in brittle systems. Note: 'N' denotes normal distribution; 'LN' denotes lognormal distribution; '0' denotes $\rho(R_i, R_j) = 0$; '0.5' denotes $\rho(R_i, R_j) = 0.5$.

application of the redundancy factor proposed by Zhu and Frangopol (2014a) without taking into consideration the effects of post-failure material behaviour. The bridge is simply supported with a span of 20 m. The deck consists of 18 cm of reinforcement concrete and an 8 cm surface layer of asphalt. The deck is supported by four I-beam steel girders whose dimensions are assumed the same. The goal of the design is to determine the bending resistance of the girders using the proposed redundancy factors taking into account the effects of post-failure material behaviour.

Because the bridge is simply supported, the maximum bending moment due to both dead loads and live loads occurs at the mid-span cross-section of the girders. The limit state equations associated with the flexure failure of girder i at the mid-span cross-section is (Zhu and Frangopol 2014a, 2014b)

$$g_i = M_{U,i} - M_{L,i} = 0, \qquad (20)$$

where $M_{U,i}$ and $M_{L,i}$ are the ultimate moment capacity and bending moments acting on girder i, respectively. Assuming the cross-section of the girder is uniform along the length, the ultimate moment capacity at mid-span cross-section, $M_{U,i}$, will govern the design. In order to find the mean value of the ultimate moment capacity of each girder, the maximum bending moments of girders due to dead and live loads need to be determined.

As shown in Zhu and Frangopol (2014a): (a) the total bending moments associated with the exterior and interior girders are $M_{L,ext} = 3407\,kN\,m$ and $M_{L,int} = 3509\,kN\,m$, respectively; (b) in the limit state (see Equation (20)), both the moment capacity and load effect are assumed to be normally distributed random variables with the coefficients of variation of $M_{U,i}$ and $M_{L,i}$ equal to 0.05 and 0.3, respectively; (c) the total bending moments are used as the mean values of the $M_{L,i}$ in Equation (20); and (d) based on statistical information on the capacity and load effect, the mean resistances of exterior and interior girder when their reliability indices are 3.5 are: $E_c(M_{U,ext}) = 7200\,kN\,m$ (exterior girder) and $E_c(M_{U,int}) = 7415\,kN\,m$ (interior girder). The larger value between the mean resistances of exterior and interior girders is used as the final mean resistance of the girders: $E_c(M_U) = 7415\,kN\,m$.

As mentioned previously, the redundancy factor associated with series system is independent of the post-failure behaviour of components. Therefore, only the parallel and series-parallel systems are considered herein. For the investigated four girders, the four-component parallel system and the 2×3 series-parallel system can be formed based on two different definitions of system failure: (a) the system fails only if all girders fail (parallel system) and (b) the system fails if any two adjacent girders fail (series-parallel system). Because the girders are made of steel which is a ductile material, the redundancy factors associated with the ductile case will be applied.

The redundancy factors η_R of the four-component ductile parallel system associated with three correlation cases are provided in Table 1. By performing the procedure for calculating the redundancy factors of ductile systems, η_R of the 2×3 series-parallel system can also be determined with respect to three correlation cases as: 1.009 if $\rho(R_i, R_j) = 0$, 1.012 if $\rho(R_i, R_j) = 0.5$ and 1.0 if $\rho (R_i, R_j) = 1.0$. By multiplying the mean resistances of girders obtained previously by these redundancy factors, the designed mean resistances of girders in parallel and series-parallel systems are obtained, as listed in Table 10.

It is noticed that (a) for both systems, the designed resistances of girders associated with the partial correlation case are higher than those associated with the no correlation case; (b) in the no correlation and partial correlation cases, the designed mean resistances of girders in the parallel system are less than those in the series-parallel system and (c) the designed mean resistances of girders associated with the perfect correlation case are the same for both systems considered.

With the designed resistances of girders, the total bending moments due to dead and live loads, and the associated statistical parameters, the reliability indices of the exterior β_{ext} and interior β_{int} girders in the parallel and series-parallel systems are evaluated, as listed in Table 10. It is found that: (a) in both systems, the reliability indices associated with interior girders are less than those associated with the exterior girders; (b) compared with the parallel system, the reliability indices of girders of series-parallel systems are higher and (c) as the correlation among resistances of girders increases, the reliability indices of girders increase in the parallel system.

6. Conclusions

This paper investigates the redundancy factors of systems considering the post-failure behaviour of components. Systems consisting of two to four components are used as examples to demonstrate the procedure for evaluating the redundancy factors of ductile, brittle and mixed systems. The effects of number of brittle components in a system and the post-failure behaviour factor on the redundancy factor are also studied using these systems. In order to generate standard tables to facilitate the design process, the redundancy factors of N-component nondeterministic ductile and brittle systems with a large number of components are calculated with respect to three correlation cases and two probability distribution types. Finally, a bridge example is studied to demonstrate the application of the ductile redundancy factor in the design of steel girders. For the systems analysed, the following conclusions are drawn:

1. An approach for simplifying the system model in the redundancy factor analysis of brittle systems is

Table 10. The designed mean resistances of girders and the reliability indices of the exterior and interior girders.

System type	Correlation case	Designed mean resistance of girders (kN m)	β_{ext}	β_{int}
Parallel system	$\rho(R_i, R_j) = 0$	7252	3.55	3.36
	$\rho(R_i, R_j) = 0.5$	7333	3.61	3.43
	$\rho(R_i, R_j) = 1.0$	7415	3.69	3.50
Series-parallel system	$\rho(R_i, R_j) = 0$	7482	3.74	3.56
	$\rho(R_i, R_j) = 0.5$	7504	3.77	3.57
	$\rho(R_i, R_j) = 1.0$	7415	3.69	3.50

proposed. By reducing the $N! \times N$ series-parallel system model to the $N \times N$ series-parallel system model, this approach makes it possible to calculate the redundancy factor of components in brittle parallel systems with a large number of components.

2. The redundancy factors associated with series, parallel and series-parallel systems in the brittle case are the same; this indicates that for an N-component brittle structure, the redundancy factor is independent of the system modelling type.

3. For the ductile parallel system consisting of only a few components, increasing N leads to a significant decrease of the redundancy factor. However, as N continues increasing, this decrease becomes insignificant.

4. Increasing the correlation among the resistances of components leads to higher redundancy factors in the ductile parallel system, respectively. The difference in the redundancy factors between the normal and lognormal distributions is more significant in the ductile parallel than in the series system.

5. The redundancy factors of the mixed parallel systems are at least 1.0 due to the existence of brittle component(s) in the systems. As the number of brittle components increases in an N-component mixed system, the redundancy factor becomes larger and closer to the redundancy factor associated with the brittle case. Increasing the correlation among the resistances of components leads to a lower redundancy factor in the mixed parallel systems.

6. As the post-failure behaviour factor increases in the no correlation and partial correlation cases, the redundancy factor of the parallel system initially remains the same and then decreases dramatically.

Acknowledgements

The support from the US Federal Highway Administration Cooperative Agreement 'Advancing Steel and Concrete Bridge Technology to Improve Infrastructure Performance' Project Award DTFH61-11-H-00027 to Lehigh University is gratefully acknowledged. The opinions and conclusions presented in this paper are those of the authors and do not necessarily reflect the views of the sponsoring organisation.

Note

1. Email: bez209@lehigh.edu

References

American Association of State Highway and Transportation Officials. (1994). *LRFD bridge design specifications* (1st ed.). Washington, DC: Author.

American Association of State Highway and Transportation Officials. (2012). *LRFD bridge design specifications* (6th ed.). Washington, DC: Author.

Ang, A.H-S., & Tang, W.H. (1984). *Probability concepts in engineering planning and design* (Vol. 2). New York, NY: Wiley.

Babu, S.G.L., & Singh, V.P. (2011). Reliability-based load and resistance factors for soil-nail walls. *Canadian Geotechnical Journal, 48*, 915–930.

Cavaco, E.S., Casas, J.R., & Neves, L.A.C. (2013). Quantifying redundancy and robustness of structures. *Proceedings of IABSE workshop on safety, failures and robustness of large structures*. International Association for Bridge and Structural Engineering (IABSE), Zurich, Switzerland.

Ellingwood, B., Galambos, T.V., MacGregor, J.G., & Cornell, C.A. (1980). *Development of a probability-based load criterion for American National Standard A58*. NBS Special Publication 577. Washington, DC: U.S. Dept of Commerce.

Frangopol, D.M., & Curley, J.P. (1987). Effects of damage and redundancy on structural reliability. *ASCE Journal of Structural Engineering, 113*, 1533–1549.

Frangopol, D.M., & Klisinski, M. (1989a). Material behavior and optimum design of structural systems. *ASCE Journal of Structural Engineering, 115*, 1054–1075.

Frangopol, D.M., & Klisinski, M. (1989b). Weight–strength–redundancy interaction in optimum design of three-dimensional brittle–ductile trusses. *Computers and Structures, 31*, 775–787.

Frangopol, D.M., & Nakib, R. (1991). Redundancy in highway bridges. *Engineering Journal, American Institute of Steel Construction, 28*, 45–50.

Fu, G., & Frangopol, D.M. (1990). Balancing weight, system reliability and redundancy in a multiobjective optimization framework. *Structural Safety, 7*, 165–175.

Ghosn, M., & Moses, F. (1998). *Redundancy in highway bridge superstructures* (NCHRP Report 406). Washington, DC: Transportation Research Board.

Ghosn, M., Moses, F., & Frangopol, D.M. (2010). Redundancy and robustness of highway bridge superstructures and substructures. *Structure and Infrastructure Engineering, 6*, 257–278.

Hendawi, S., & Frangopol, D.M. (1994). System reliability and redundancy in structural design and evaluation. *Structural Safety, 16*, 47–71.

Hsiao, L., Yu, W., & Galambos, T. (1990). AISI LRFD method for cold-Formed steel structural members. *Journal of Structural Engineering, 116*, 500–517.

Kim, J. (2010). *Finite element modeling of twin steel box-girder bridges for redundancy evaluation* (Dissertation). The University of Texas at Austin, Austin, TX.

Lin, S., Yu, W., & Galambos, T. (1992). ASCE LRFD method for stainless steel structures. *Journal of Structural Engineering, 118*, 1056–1070.

Liu, D., Ghosn, M., Moses, F., & Neuenhoffer, A. (2001). *Redundancy in highway bridge substructures* (National Cooperative Highway Research Program, NCHRP Report 458, Transportation Research Board). Washington, DC: National Academy Press.

MathWorks. (2009). *Statistical toolbox*. MATLAB Version 7.9. The MathWorks Inc., Natick, MA.

Okasha, N.M., & Frangopol, D.M. (2010). Time-variant redundancy of structural systems. *Structure and Infrastructure Engineering: Maintenance, Management, Life-Cycle Design and Performance, 6*, 279–301. doi:10.1080/15732470802664514

Paikowsky, S.G. (2004). *Load and resistance factor design (LRFD) for deep foundations* (NCHRP Report 507). Washington, DC: Transportation Research Board.

Rabi, S., Karamchandani, A., & Cornell, C.A. (1989). Study of redundancy of near-ideal parallel structural systems. *Proceedings of the 5th international conference on structural safety and reliability*. pp. 975–982. ASCE: New York, NY.

Thoft-Christensen, P., & Baker, M.J. (1982). *Structural reliability theory and its applications*. Berlin: Springer-Verlag.

Thoft-Christensen, P., & Murotsu, Y. (1986). *Application of structural systems reliability theory*. Berlin: Springer-Verlag.

Tsopelas, P., & Husain, M. (2004). Measures of structural redundancy in reinforced concrete buildings. II: Redundancy response modification factor RR. *Journal of Structural Engineering, 130*, 1659–1666.

Wen, Y.K., & Song, S.-H. (2004). Structural reliability/redundancy under earthquakes. *Journal of Structural Engineering, 129*, 56–67.

Zhu, B., & Frangopol, D.M. (2014a). Redundancy-based design in nondeterministic systems. In D.M. Frangopol & Y. Tsompanakis (Eds.), *Safety and maintenance of aging infrastructure*, Chapter 23. Boca Raton, FL: CRC Press. doi:10.1201/b17073-24 (in press).

Zhu, B., & Frangopol, D.M. (2014b). Effects of postfailure material behavior on the reliability of systems. *ASCE-ASME Journal of Risk and Uncertainty in Engineering Systems, Part A: Civil Engineering*, doi:10.1061/AJRUA6.0000808 (in press).

Design and construction of two integral bridges for the runway of Milan Malpensa Airport

Pier Giorgio Malerba and Giacomo Comaita

In this study, two bridges recently completed at the Malpensa Airport are presented. The bridges overpass the trench where the railway track from Malpensa Airport runs to the city of Milan. They are part of a U-shaped path, called the hotel (H) ring, which connects the southern ends of the two existing taxiways. Such a crossing had to: (a) intersect the railway track twice with a high skew angle, (b) carry the maximum aeronautical loads, corresponding to the gears of an AIRBUS 380 aircraft and (c) host a 60-m wide taxiway strip, plus a lateral safety lane and a service road, for a total width of 96 m. For a bridge operating in such an environment, it was imperative to reduce inspections and maintenance activities to a minimum. Therefore, the designers excluded any detail arrangements or auxiliary devices having life-cycle performances much shorter than that of the main structure. In particular, expansion joints and movable or deformable bearing supports were excluded. Such constraints led to design two structures having the peculiar features that characterise them as integral bridges. This paper will firstly recall the aeronautical and railway specifications and the other operating constraints that characterised the design. In the second part, the criteria followed in designing the foundations, the abutments and the deck are presented. The structural analyses were set out in accordance to the sequence of the construction stages. For the bridge in its service configuration, self-weight and permanent loads give rise to a set of internal forces typical of a portal frame, while the localised loads, due to the aircraft gears, induce bubble deformed shapes characterised by particular transversal load distribution mechanisms. Details of completions and finishing works are finally given.

Introduction

The construction of the Hotel (H) ring is part of a general project to promote Malpensa Airport as an international hub. The project started in the 1990s and included the construction of a new terminal, a freight area called Cargo City and a new apron capable of hosting more than 100 aircrafts at one time. A new road network, including two landmark cable-stayed bridges, a railway station, a multi-storey car park and a new hangar for Category E aircrafts were also built. The project was completed in October 2010 with the opening of a large hotel in the area opposite the air terminal. The construction of the H ring involved overpassing the trench which has a railway track from Malpensa Airport to the city of Milan. The H ring is a U-shaped path, which connects the southern ends of the two existing taxiways and intersects the railway track twice, with a skew angle of about 45°. The intersection was solved by means of two bridges astride the railway (Figure 1).

There are bridges that are meant to become landmarks; they impose themselves on the environment, so that the introduction of a new infrastructure involves the transformation of the surrounding landscape into a different skyline. In contrast, there are bridges, which are subjected to so many functional and geometrical constraints that their conception turns out simply as the best compromise among the limits imposed by different needs. This is just the case of the twin runway bridges of Milano Malpensa Airport: while they respond to a high number of strict specifications of different origins, their footprint on the surrounding landscape is negligible as they are practically invisible.

Such bridges have many peculiarities. The first one is that their spans are relatively small ($l = 21.50$ m for the West bridge and $l = 19.50$ m for East bridge), while the decks are very wide, so that they can host a 60-m wide taxiway strip, plus a lateral safety lane and a service road, for a total width of 105 m. Second, they have to carry the huge and very localised loads corresponding to the gears of an AIRBUS 380 aircraft, at present the maximum aeronautical loads. Furthermore, in order to limit any interference with both the airport and the railway activities, they must require little inspection and minimum maintenance.

The request for a great durability and low-maintenance costs oriented the designers to a so-called integral construction or, in other words, to design bridges shaped as portal frames, in which the deck is continuous and connected monolithically with the abutments walls, with no expansion joints or movable or deformable bearing

Figure 1. Aerial view of the Hotel taxiway ring.

supports (Burke, 2009). On one hand, the portal shape makes the internal force distribution more effective and possesses superior seismic performance, thanks to the frame action and the arching geometry (Hambly, 1991; Masrilayanti & Weekes, 2012). On the other hand, it is known that continuous structures are also very sensitive to the effects of compressibility of supports and of settlements of foundations. For this reason, the analyses paid great attention to the soil abutment interaction and several models and design criteria were explored (Lehane, Keogh, & O'Brien, 1999).

In the typical integral bridge, the completion of the abutments is followed by the construction of the girder and by the backfilling behind the abutments. In this way, the backfilling presses against the stems of a closed portal frame and the internal forces on the abutments are those of a system fully or partially restrained at its ends. In the present case, such a sequence of phases was not possible. In particular it was not possible a direct construction of the deck, and it was not possible to lay heavy precast prefabricated beams by using a crane positioned behind the ridges of the excavation, which were too far with respect to the vertical plan of the abutments. For this reason, backfilling preceded the closure of the frame, and the abutments started their life working as earth-retaining walls, clamped at bases and free at the top. Therefore, the design of these temporary phases conditioned the dimensioning of the vertical walls.

Unlike the standard integral bridge, for which the abutments are simple vertical walls, in this case, due to the unusual width of the deck (146 and 134 m respectively), the abutments were conceived as an array of diaphragms, fixed to the continuous foundations and clamped to the beams that transversally connect the precast girders. These diaphragms are separated by vertical joints in order to diminish the shear action that would arise from possible differential settlements in case the walls were continuous.

For the sake of durability, great care has been paid to the study of the details that had to be appropriate for the importance and the dimensions of the bridges. Among

these, the dimensioning of the concrete covers, the setup of an effective continuous drainage system behind the abutment walls and the adoption of segmented approach slabs at the two ends of the bridges, in order to provide a smooth transition between the approach pavement and the rigid superstructures, are worth mentioning.

In the sequence, the aeronautical and railway specifications and the other operating constraints that characterised this crossing are recalled. Then the criteria followed in designing the foundations, the abutments and the deck are presented. The structural analyses were set out in accordance with the sequence of the construction stages. For the bridge in its service configuration, self-weight and permanent loads give rise to nearly cylindrical flexure and therefore to the set of internal forces typical of a portal frame, while the load due to the aircraft gears, being exerted on very localised areas, induce bubble deformed shapes and activate particular transversal load distribution mechanisms. Finally, details of completions and finishing works are given.

Aeronautical specifications

The runway characteristics, the radius of the bends and the elevation of the platform were prescribed by aeronautical specifications. The runways were to be designed for taxying aircrafts up to Airbus 380 class. The wing span of such an aircraft is 79.75 m, its length is 72.72 m and the total height at top of the tail is 24.09 m (Figure 2). These specifications resulted in a very large taxiway strip, whose width is that of the runway itself (60 m for this class of aircrafts), plus that of a lateral safety edge strip (30 m) and that of a lateral service road (6 m), for a total of 96 m.

One of the characterising features of these bridges is the magnitude of the service loads, on account of the transit of AIRBUS 380/843F–863F. The load intensities are summarised in Figure 3. In order to allow rainwater run-off, the bridge platform was shaped as an inclined plan, sloping in both the longitudinal and the transversal directions. The elevation differences with respect to the bare extrados of the deck were achieved by means of a concrete layer of varying thickness. This implied a varying dead load on the deck. Furthermore, airport maintenance specifications required no joints, no mechanical bearing supports and effectiveness of waterproofing and drainage systems, in order to avoid both frequent inspections and expensive maintenance operations. In other words, high durability for all internal and exposed parts was required.

Railway specifications

The bridges overpass the trench which has a railway track linking the Malpensa Airport to the city of Milan. Both bridges intersect the railway track with a high skew angle.

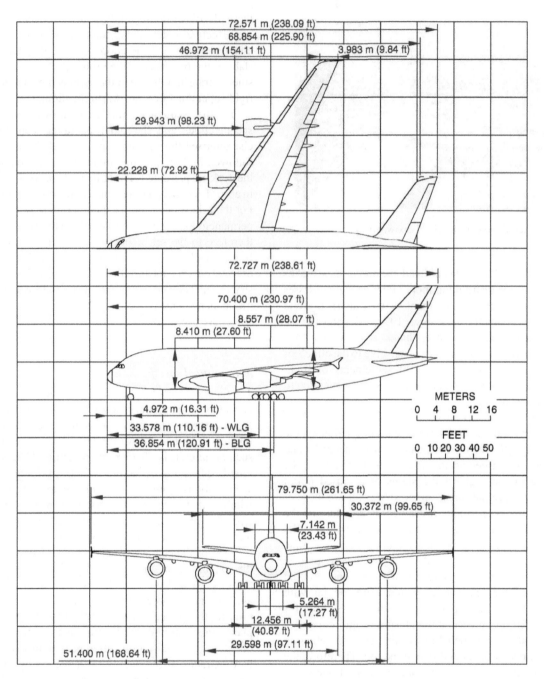

Figure 2. Main dimensions of an AIRBUS 380 aircraft.

The area enclosed by the bridges was determined by the underlying rail platform, which consists of three tracks (two operating, one under construction) and a service road used for inspection and maintenance operations, for a total width of about 21 m (Figures 4 and 5).

The aspect of the railway clearance required the bridge intrados to be at least 6 m above the track surface and the distance between the bridge intrados and the live power line to be no less than 1.5 m. The extrados of the foundation had to be at least 0.50 m below the base of the railway platform, hence at a level of 9.5–10.50 m below

the reference ground altitude. The constraints imposed by the presence of the railway were not only of geometrical nature. In fact, during construction, the railway was fully operating from 5 a.m. until 9 p.m., with trains passing at a maximum speed of 140 km/h. Hence, no interferences were permitted and special operations over the railway, such as the laying of the bridges girder beams, had to be performed at night only.

A specifically designed fence, able to withstand the air overpressure due to the abovementioned maximum train speed, was installed in order to keep the railway platform

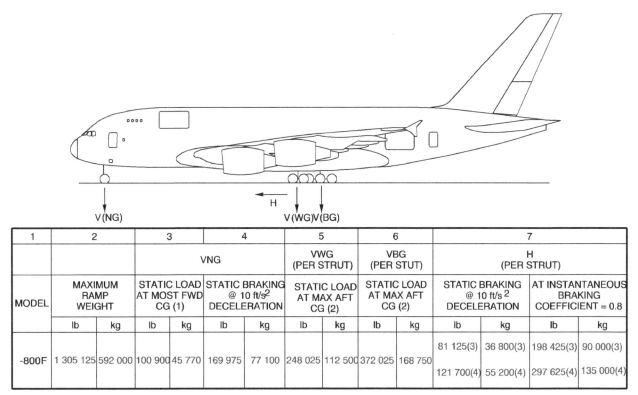

1	2		3		4		5		6		7			
			VNG				VWG (PER STRUT)		VBG (PER STUT)		H (PER STRUT)			
MODEL	MAXIMUM RAMP WEIGHT		STATIC LOAD AT MOST FWD CG (1)		STATIC BRAKING @ 10 ft/s^2 DECELERATION		STATIC LOAD AT MAX AFT CG (2)		STATIC LOAD AT MAX AFT CG (2)		STATIC BRAKING @ 10 ft/s^2 DECELERATION		AT INSTANTANEOUS BRAKING COEFFICIENT = 0.8	
	lb	kg	lb	kg	lb	kg	lb	kg	lb	kg	lb	kg	lb	kg
-800F	1 305 125	592 000	100 900	45 770	169 975	77 100	248 025	112 500	372 025	168 750	81 125(3) / 121 700(4)	36 800(3) / 55 200(4)	198 425(3) / 297 625(4)	90 000(3) / 135 000(4)

V(NG)   MAXIMUM VERTICAL NOSE GEAR GROUND LOAD AT MOST FORWARD CG

V(WG)   MAXIMUM VERTICAL WING GEAR GROUND LOAD AT MOST AFT CG

V(BG)   MAXIMUM VERTICAL BODY GEAR GROUND LOAD AT MOST AFT CG

H   MAXIMUM HORIZONTAL GROUND LOAD FROM BRAKING

(1)   FWD CG = 36.0 % MAC

(2)   AFT CG = 42.8 % MAC

(3)   BRAKED WING GEAR

(4)   BRAKED BODY GEAR

NOTE: ALL LOADS CALCULATED USING AIRPLANE MAXIMUM RAMP WEIGHT

Figure 3.   Loads transmitted by the landing gears of AIRBUS 380/843F–863F aircrafts.

physically separated from the working area (Figure 6). As concerns the deck construction, the use of a large and heavy scaffolding over an operation railway for a cast-in-place solution was not a viable option. The only possible choice, which was also the most rational one, was the use of precast, prefabricated beams subsequently completed by cast-in-place connecting transversal beams and by the upper slab.

The need of working in the presence of live power cables represented a significant safety issue. The position of the hanging points of the power line at the intrados of the deck had to be moved so that in every given stage of construction, the resulting catenary did not interfere with the structure: each time a repositioning of the hanging points was required, this had to be completed in one night so as to allow for the line to be operating the following day.

Figure 4.   The West Bridge under construction.

Other constraints and soil characteristics

The elimination of all joints and bearing supports imposed by the airport authorities meant that the structure could only be shaped as portal frame, in which the deck is

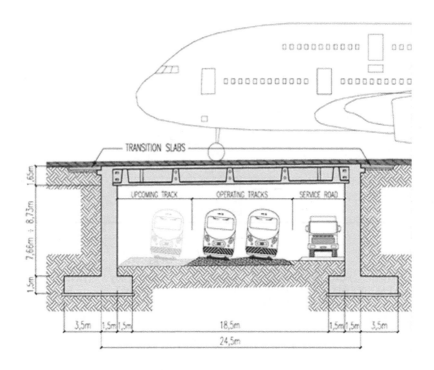

Figure 5. Cross section of the west bridge. The relative proportions among bridge, aircraft and railway arrangement can be appreciated.

continuous and connected monolithically with the abutment walls. In structures like these, the foundation's role is not limited to transmitting the loads to the soil because the flexural stiffness of the walls influences the overall behaviour of the portal. Different types of foundations have been examined, but the final choice was driven by two major factors: the soil characteristics and the narrowness of the operating area.

The soil at the airport location is of the fluvioglacial type, whose stratigraphy is characterised by layers of weak

Figure 6. Temporary fence, designed for a maximum transit speed of 140 km/h, foundation slab and continuity reinforcement.

silty sand and gravel, covered by a layer of anthropic deposits whose thickness varies from 1 to 3 m (Figure 7). The foundation level was to be set at 11–12 m below the present ground level. At that depth, the soil resulted mainly composed of pebbles having 5–10 cm diameter. The water table was well below the foundation level. From the operational point of view, the digging operations had to be limited to the strip defined by the base of the railway superstructure and the base of the existing lateral slope of the cutting. Large excavations or the use of drilling machines would have been problematic.

From the conceptual design to the layout of the actual bridges

Bridge geometry

The two bridges are located to the west (towards Malpensa) and to the east (towards Milan) of the longitudinal axis of the main runway (Figures 1 and 8). The bridges decks have straight axes intersecting the curved railway track at a skew angle of about 45°, and are made of pre-tensioned, precast beams, made continuous by cast-in-place transverse beams and upper slab. The geometry of these structures differs from that of a conventional bridge, whose length is usually greater than its width. In this case, in fact, the West Bridge has a 146-m wide deck and a 21.50-m long effective span, while the East Bridge has a 134-m wide deck and a 19.50-m long

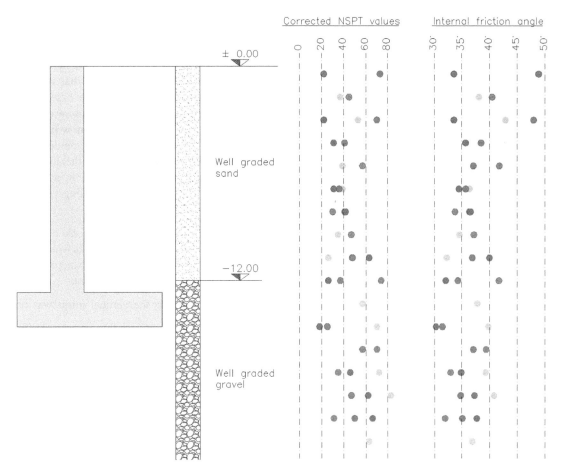

Figure 7.   Soil stratigraphy and standard penetration test (SPT) results at various depths.

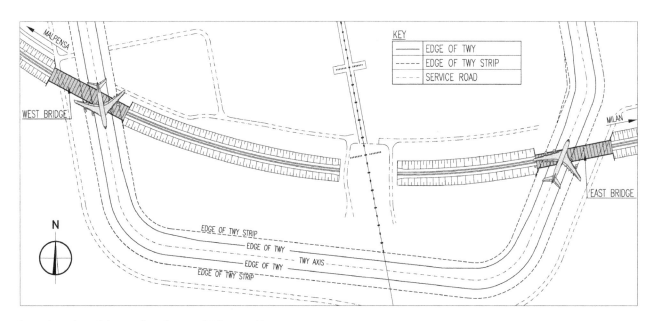

Figure 8.   Plan of the south taxiway at Malpensa Airport.

Figure 9. Diaphragms composing the backwalls. The fence, the formworks, the rebars and of the joint waterstops can be seen.

effective span. The cross section of the West Bridge is shown in Figure 5.

The vertical walls supporting the decks have variable height and also serve as retaining walls for the embankments on the two sides of the trench. The two bridges are supported on ribbon spread footings. For all the aforementioned characteristics (structure shaped like a portal frame, with abutments working as retaining walls and without bearing supports and joints), these bridges can be properly classified as integral bridges.

Foundations

It has already been mentioned that the limited operational area between the railway track and the slopes of the trench would have made it difficult to build deep foundations. Due to the reasonably good bearing capacity of the soil, the best choice seemed to support the two bridges on spread footings, running parallel to the chords of the curved railway track for the whole width of the bridges. The base of the foundation is at 196.82 m above sea level and the railway platforms are at 199.97 m (West Bridge) and 200.04 m (East Bridge). The spread footing is 6.50 m wide, with a 1.50-m wide toe and a heel protruding for 3.50-m beneath the embankment.

Abutments

The backwalls are 1.50 m thick and have a height varying from 7.66 to 8.73 m (Figure 5). A variable height is required because the deck is made of precast beams laid on an inclined plane, so as to create the slope required for rainwater run-off. The backwalls are conceived as an array of diaphragms, 1.50 m thick and 8.00 m wide, flanked adjacent to each other for the entire length of the deck

(Figure 9). At the base, they are fixed to the foundations, while at the top they are clamped to the end beams that transversally connect the precast girders.

These diaphragms are separated by vertical joints in order to mitigate the shear action that would arise from possible differential settlements in case the walls were continuous. In fact, in continuous walls, the flexural stiffness in the transverse direction is that of the whole section, while with the adopted segmentation, the inertial coupling between the upper and the lower part of the structure is avoided, thus resulting in a reduced flexural stiffness. The advantages of providing vertical joints can be summarised as follows:

- more flexibility in the transverse direction;
- reduction of the onset of cracks which tend to appear on continuous walls due to concrete shrinkage and delayed differential settlement.

Waterproofing across the vertical joints was provided by waterstop seals.

The deck

Runways axes and transit loads are not perpendicular to the chords of the curved railway track. Hence the bridges behave as skew structures. At the conceptual design phase, two alternatives were possible:

- align the beams with the axis of the taxiways, which would have resulted in the beams being aligned to the abutments at a high skew angle;
- align the beams perpendicular to the abutments.

From a geometrical point of view, the second option creates two useless, although limited, triangular areas at the sides of each bridge (Figure 9); on the other hand, it brings along a series of advantages, first of all, the simplification of the erection phases and of the launching of precast beams.

There are also significant structural reasons to prefer the second option. As well known, skew girders react to imposed loads through complex combined bending-torsion regimes and tend to overload the obtuse angles and to unload the others two. From an analytical point of view, nowadays this is no longer a problem. However, problems still remain when the analysis results have to be translated into a rational and feasible reinforcement layout. Concerning this issue, one can observe that the flow of the principal flexural moments tends to align with the least path between the support lines, i.e. with the minor diagonal. On the other hand, when the transverse width ($2b$) of the deck is much higher than the span (l) (in this case $2b/l \cong 5-6$), the least path between the support lines tends to be perpendicular to the abutments. Therefore, by placing the beams perpendicular to the abutments, a rectangular deck of least span is obtained. The transversal end beams, cast on the walls,

Figure 10. Precast beam prior to launching. On the right, the waterproofed wall made of diaphragms and the continuity reinforcement can be seen.

Figure 11. Precast beam at the manufacturing plant. The reinforcement and one end anchor plate with the set of strands before tensioning can be seen.

provide a regularisation action and call for all the possible structural collaborations.

According to this preliminary concepts, the West Bridge deck comprises of 73 precast, prestressed beams, having a 22.30-m span. The East Bridge comprises 67 beams, spanning 20.30 m. All beams have a constant depth of 1.65 m. The beams procured by the contractor slightly differ from the U beams considered in the design, on account of two lateral appendixes protruding at the bottom (Figure 10).

The precast girders lean for a length of 0.40 m on the backwalls of the abutments, whose overall width is 1.50 m. The $(1.50 - 0.40) = 1.10$-m gaps left on either sides, between the end of the precast girders and the embankment, were filled in a second phase, after the casting of the slab. Three more similar transversal beams subdivide the span. The structure is completed by the cast-in-place upper slab, whose thickness varies from 0.25 to 0.49 m. The overall behaviour of the platforms is, thus, similar to that of a cellular deck, characterised by strong transverse connections and by high flexural and torsional stiffness.

Once the upper slab was cast, the deck and the abutments were made continuous. The resulting structure behaves as a portal. Its stiffness is greater than that of a simply supported deck and any required provision for movement in the carriageway is the placed outside the structure. The precast beams are prestressed by means of 44 (West Bridge) and 40 (East Bridge) strands (Figure 11). The three transverse beams in the deck span are prestressed by two sets of cables composed of $12\varnothing0.6''$ stabilised strands. The characteristic strength of the concrete is 55 N/mm^2 for the precast beams and other elements of superstructure and 35 N/mm^2 for the abutments and the foundations. An average concrete cover of 50 mm was adopted for all members.

Construction stages and their influence on the analysis of the structure

Given the presence of the operating railway, the first stage of the work consisted of making the construction site safe by building two lateral fences, 6.0 m height, positioned on both sides at a distance of about 2.5 m from the external rails. The fences, which were meant to protect the operating railway against possible debris coming from the working area, had to withstand the pressure exerted by the transit of the trains at a design speed of 140 km/h. This resulted in a very high maximum pressure value ($q = 0.25$ kN/m^2), applied on very large surfaces. The fences sheets were therefore supported by means of strong vertical H-section European Light series 120 columns, each founded on a single $\varnothing$273-mm pile driven 4.0 m into the ground. In order to reduce the forces on columns and piers, the surfaces of the fences were made partially by solid wood panels and partially by plastic perforated sheets (Figure 6).

The following phases concerned the excavation of the two lateral trenches, the placement of the formworks and of the reinforcement and the casting of the footings (Figure 6). The footings had to be longitudinally continuous, so at each stage of the casting, rebar segments were left in order to maintain the reinforcement continuity. In a similar way, vertical rebar segments were left in order to provide continuity with the vertical walls. Once the footings had been completed, the abutments were cast. A bituminous waterproofing layer was applied on the basement and on the vertical surfaces of the walls. Moreover, on the vertical walls, a continuous drainage curtain was built by means of 22-mm thick panels, put side by side, each made of two membranes, imbricated to each other through plastic filaments. The external surfaces of these membranes were made of geotextile, meant to prevent fine particles from filtering into the drainage wall and obstructing it.

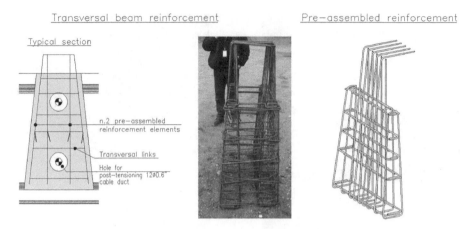

Figure 12. Head section of a transverse beam showing the position of the postensioning cables; transverse beam preassembled reinforcement.

In an integral bridge, the completion of the abutments is usually followed by the construction of the girder and by the backfilling behind the abutments. In this way, the backfilling presses against the stems of a closed portal frame and the internal forces on the abutments are those of a system fully or partially restrained at its ends. In the present case, such a construction sequence was not possible: a cast-in-place deck was not viable, as no scaffolding were allowed, and using a crane positioned behind the ridge of the excavation to lay heavy precast beams was equally impossible. Therefore, backfilling preceded the closure of the frame and the hearth pressure had to be supported directly by the abutments, working as earth-retaining walls, clamped at bases and free at the top. Obviously, these temporary phases had to be taken into consideration for the dimensioning of the vertical walls.

The backfilling was made of 0.50-m thick, mechanically compacted layers. When the backfilling reached the soil elevation, it was possible to lay the 55-t heavy prefabricated beams, operating from the edge of the cutting on the south side by means of cranes cantilevering up to 18–20 m and with a maximum lifting capacity in a nearly vertical position of 500 t. The precast prefabricated beams were manufactured and pretensioned in a factory about 115 km far from the construction site, and moved at night with special trucks to the airport. The beams have a channel section. The transverse beams, the upper slab and the end ribs of connection with the abutment had to be cast in different consecutive phases.

First, the three internal transversal beams were reinforced with a special preformed arrangement of bars and then cast (Figure 12). After 14 days curing period, they were postensioned. Postensioning preceded the completion of the deck in order not to disperse its transversal compaction effects on the array of the beams and allow an easier adjustment of the ends of the beams on the supports at the top of the abutments. To this purpose, 100×100-mm neoprene plates were placed all along the surface of laying, in correspondence to the axes of the beam webs. Subsequently, the reinforcement bars of the upper slab and of the end ribs, together with those at the corner connecting the deck and abutments, were positioned. Finally, the cast completed the main part of construction.

Structural analysis

The structural analysis was carried out on the schemes emerging from the preliminary design and had to follow the evolution of the structure during the construction stages and the changes of the static schemes and of the applied loads. A first set of analyses concerned the lateral abutments alone, without the upper deck. The second set studied the portal frame, subjected to the added permanent loads, to the traffic loads and to temperature effects.

Abutments

In their initial configuration, the abutments had to work as earth-retaining walls, subjected to their self weight and to the weight and lateral pressure exerted by the backfill. The structure is a simple cantilever, clamped at the base and free at the top. The earth at-rest pressure was defined by assuming a friction angle $f = 30°$ and a null cohesion. The results of this set of analyses determined the thickness and the main reinforcement of the walls.

In this phase, particular analyses concerned the steel columns bearing the fences and their supporting piles, driven into the soil. Such piles were considered also as a lateral confinement system of the railway platform, against possible settlements towards the excavations at the two sides (Figure 13). Their contribution was crucial in the rainy weather, when the water tended to wash away the soil on the lateral slopes.

Figure 13.   Temporary fence foundation piles.

Portal frame analysis

In the second stage, the final static scheme was studied. The structure is shaped as an inverted channel, having a significant degree of skewness with respect to its axis and subjected to heavy localised loads due to the aircraft gears. A finite element model, made of grids of equivalent beams (Hambly, 1991; Hrennikoff, 1941; Yettram & Husain, 1965) has been used. Such a criterion was preferred with respect to other choices (combination of plates/shells elements or use of solid elements) for many reasons. In the elastic range, a beam element is defined equivalent to a continuum element if, for a given set of displacements applied at its nodes, it assumes the same strain energy of the corresponding element of continuum, subjected to the same nodal displacements.

As demonstrated by Hrennikoff, the structural behaviour is correctly represented and the global results are the same as those given by more refined theories. Most of all, by using beam/grillage models, the results are given in terms of internals forces, that is, they are directly given in terms of the integral quantities which are at the basis of the reinforced concrete theory. Two sets of analysis were carried out, according the two extreme hypotheses of perfectly hinged and of fully restrained wall basement.

Loading conditions

Special attention has been paid to modelling the load combinations, which consist of: (a) added permanent loads (runway platform), (b) applied traffic loads, (c) soil pressure effects against the abutment walls and (d) temperature effects.

Traffic loads

As concerns the applied traffic loads, many wheel positions on the deck were possible and they were systematically studied (Figure 14). The main loading conditions that were analysed are summarised in Table 1. The landing gear loads, for which a dynamic amplification factor of $\phi = 1.3$ have been adopted, are listed in Figure 3. The same positions were adopted for studying the horizontal loads

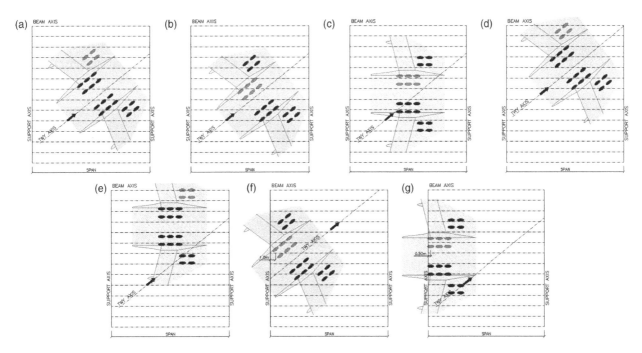

Figure 14.   The bridges were studied for different positions of the aircraft.

Table 1. Loading conditions.

Loading condition	Airplane direction	WG position	BG position	Corresponding figure
1	Aligned with TWY	At midspan of an internal beam	–	Figure 14(a)
2	Aligned with TWY	–	At midspan of a central beam	Figure 14(b)
3	Aligned with beam axis	–	At midspan of a central beam	Figure 14(c)
4	Aligned with TWY	Closest position to bridge edge	–	Figure 14(d)
5	Aligned with beam axis	–	Closest position to bridge edge	Figure 14(e)
6	Aligned with TWY	–	Closest position to support line	Figure 14(f)
7	Aligned with beam axis	–	Closest position to support line	Figure 14(g)

Note: TWY, taxiway; WG, wing gear (1125.0 kN); BG, body gear (1687.5 kN).

given by the static breaking deceleration and by instantaneous braking, whose intensities are again listed in Figure 3.

With respect to the structural behaviour, the self weight and the permanent loads give rise to nearly cylindrical flexure and, as a consequence, to axial forces, shear forces and bending moments typical of a portal frame. The loads due to the aircraft gears are exerted on very localised areas, thus inducing bubble deformed shapes like those shown in Figure 15. Figure 16 presents the normalised transversal load distribution of the shear forces for the loading conditions (c) and (e).

Soil pressure

In the final static scheme, the at-rest pressure $k_0 = k_0 (30°)$ of the soil was assumed as the one transmitted by the backfill and considered as an applied load. In order to establish a comparison with the results of such assumptions, other soil pressure models, such as those suggested by England, Tsang, and Bush (2000) and Springman and Norrish (1996), have been explored. For $f = 30°$, Figure 17 depicts the active, at-rest and passive lateral earth pressure coefficients k_a, k_0, k_p and those obtained from the theories of England (E) and Springman (S), and Figure 18 shows the corresponding lateral earth pressure distributions along the height. The effects of the different lateral soil pressure distributions and of the two end restrain conditions at the footing (fixed and hinged) are summarised with reference to a slice of bridge of unit length, behaving like a portal frame. Figures 19 and 20 illustrate the bending moments and the shear forces distributions along the wall.

With the exclusion of the hypothesis of Springman (dashed lines), all the other solutions are included in the shaded area, which is delimited by the cantilever solution $k_0(C)$ and the hinged solution with the pressure according to England E(H). The hinged solution with k_0 has a maximum bending moment a little smaller than E(H). At the end, the wall was symmetrically reinforced adopting the maximum at the base. Similar consideration may be made for the shear forces. No practical differences of the deck internal forces were measured. Hence, the final results, referred to the most important sections of both the vertical walls and the deck, were similar. Among all the solutions, the one according to the Springman pressure distribution seemed less suitable for this type of bridge because of the high stiffness of walls, deck and foundations and due to the limited temperature excursions.

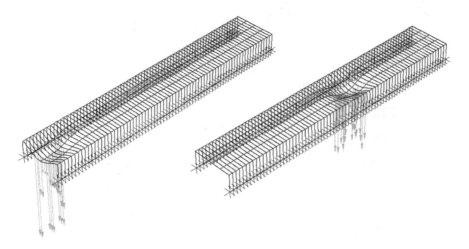

Figure 15. Deformed configuration for the loading conditions shown in Figure 14(c) and (e) (load close to the border beams).

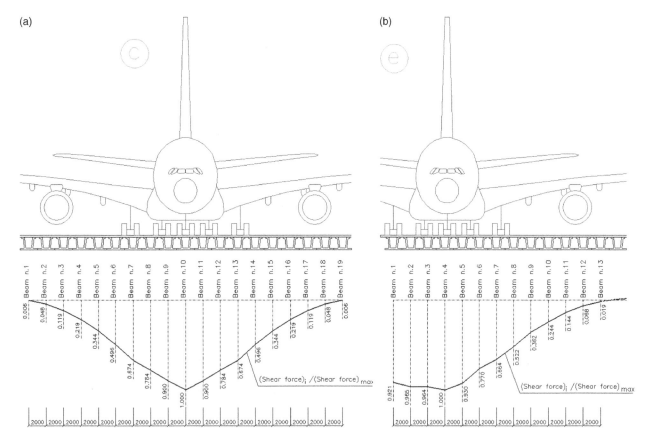

Figure 16. Normalised shear forces transversal distribution along the middle section for the loading conditions shown (a) in Figure 14(c) and (b) Figure 14(e) (load close to the border beams).

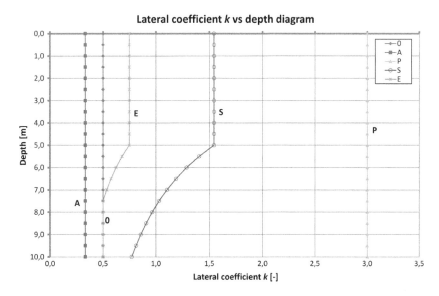

Figure 17. Earth pressure coefficients distribution along the height: k_a (active), k_0 (at-rest), k_p (passive), E (England), S (Springman). Friction angle $\varphi = 30°$.

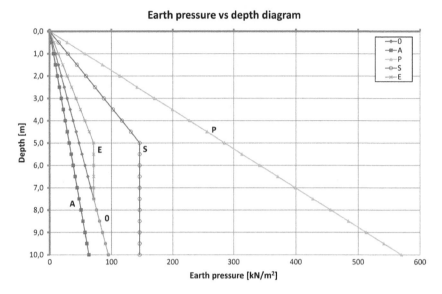

Figure 18. Lateral earth pressure distributions along the height, corresponding to the active, at-rest and passive lateral earth pressure and according to England (E) and Springman (S). Friction angle $\varphi = 30°$.

Temperature effects

According to the Italian National Code in force at the design time, the effects of the following temperature gradient have been analysed and then combined with those deriving from the loading conditions (Table 2). The movements calculated for the thermal effects were quite small (maximum lateral displacement $\delta = \pm 1.07$ mm): in service, they are compensated by slight adjustments of the soil backing the wall.

Completions and finishing works

Some finishing works followed the completion of the main structures, namely the protection of the deck surface by a continuous waterproofing made of polyester sheets, the casting of the concrete layers giving the expected longitudinal and lateral slopes, the construction of the approach slabs and of the runway bituminous surface. The main aspects of such works will be briefly presented in the following.

Approach slabs

The bridge stiffness with regards to the vertical loads is much greater than that of the platform of the taxiway. In order to provide a smooth transition, between the approach pavement and the rigid bridge superstructure, approach slabs were introduced (Taylor, 1993). In the past,

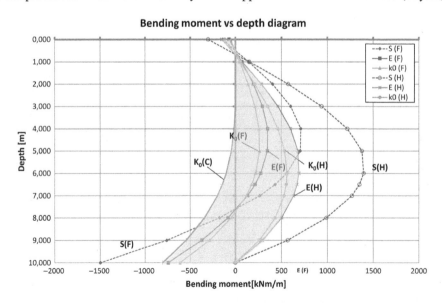

Figure 19. Bending moments along the lateral wall for different lateral earth pressure distributions and different restraints at the base (F = fixed; H = hinged and C = cantilever fixed at the base and free at the top).

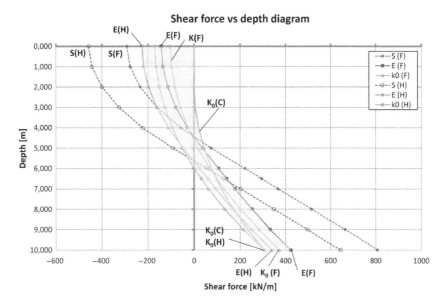

Figure 20. Shear forces moments along the lateral wall for different lateral earth pressure distributions and different restraints at the base (F = fixed; H = hinged and C = cantilever fixed at the base and free at the top).

Table 2. Thermal load conditions.

	Gradient 1	Gradient 2	Gradient 3	Gradient 4
ΔT sup	10°	10°	− 10°	− 10°
ΔT inf	0°	10°	− 10°	0°

problems arose due to poor detailing, to the failure of the connection between approach slab and abutment or to differential settlement of the backfill supporting the approach slab (Soubry, 2001). To prevent such drawbacks, each approach slab was made of an array of 300-m wide separated slabs, having an end hinged to the backwall and resting on a protruding corbel, so as to allow for tilting and remaining anchored to the wall (Figure 21). The array of slabs is laid on 200 mm sub-base, made of gravel mixed with cement and cast over a mechanically consolidated backfill.

In the design stage, the slabs were studied using the Westergaard solution of the plates on elastic soil (Westergaard, 1948). The analyses have been carried out for the cases of a concentrated load transmitted by a single wheel and positioned at the centre of the slab, at the centre of an edge and at a corner.

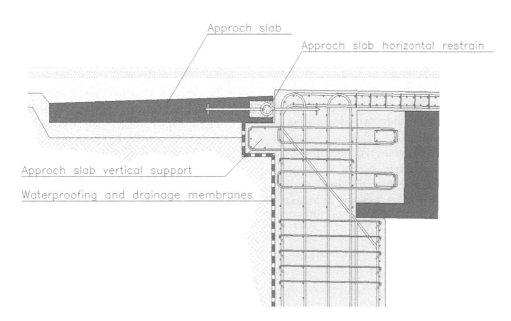

Figure 21. Approach slab and its connection to the backwall.

Table 3. Construction data.

	West Bridge	East Bridge	Total
Reinforcing steel (t)	1250	1050	2300
Prestressing steel (t)	100	90	190
Concrete (m^3)	11,000	10,000	21,000
Total cost (EUR millions)			6.578

Table 4. Dates and other relevant data.

Construction start date	June 2007
Construction completion date	May 2009
Loading tests	July 2009
Bridges opening	June 2009
Commissioner and owner	SEA S.p.A. (Milano Airport Managing Society) and Italian Ministry of Transportation
Contractor	C.I.C. S.p.A., Varese, Italy
Beams manufacturer	CODELFA S.p.a., Tortona, Italy

Platform superstructure

The platform superstructure was composed of four layers: (1) a varying depth layer of fibre-reinforced cemented aggregate (10–31 cm), (2) a base layer made of bituminous aggregate (20 cm), (3) a fibre-reinforced binder (8 cm), (4) a bituminous wering course (6 cm).

Conclusions

The two runway bridges overpassing the railway linking the Malpensa Airport to the city of Milan have been presented. They are part of a U-shaped path which connects the southern ends of the two existing taxiways and had to: (a) intersect the railway track twice with a skew angle of about 45°, (b) carry the aeronautical loads corresponding to the gears of an AIRBUS 380 aircraft and (c) host a 60-m wide taxiway strip, plus a lateral safety lane and a service road, for a total width of 96 m.

The paper presents how the constraints given by the aeronautical, railway and airport maintenance specifications led to the design of a structure shaped like a portal frame, with abutments working as retaining walls and without bearing supports and joints. For the aforementioned characteristics, these bridges can be properly classified as integral bridges, which in this case are adopted in a peculiar context, characterised by a severe geometry constraints and an unusual magnitude of service

loads. Table 3 summarises the quantities of materials employed for construction; Table 4 provides information about the duration of the works and other relevant data regarding the two bridges.

Acknowledgements

The authors wish to acknowledge the cooperation received from the staff of the New Buildings and Plants Department of SEA S.p. A., and in particular from its Directors, Renzo Gorini, Gianpaolo Pirani and Giordano Paracchini.

Note

1. Email: mail@studiomalerba.net

References

Burke, M.P. Jr (2009). *Integral and semi-integral bridges.* Chichester: Wiley-Blackwell, John Wiley & Sons.

England, G.L., Tsang, N.C.M., & Bush, D.I. (2000). *Integral bridges: A fundamental approach to the time–temperature loading problem.* London: Thomas Telford.

Hambly, E.C. (1991). *Bridge deck behaviour.* London: E & FN Spon, Taylor and Francis Group.

Hrennikoff, A. (1941). Solution of problems in elasticity by the framework method. *Journal of Applied Mechanics, 8,* 169–175.

Lehane, B.M., Keogh, D.L., & O'Brien, E.J. (1999). Simplified elastic model for restraining effects of backfill soil on integral bridges. *Computers and Structures, 73,* 303–313.

Masrilayanti, M., & Weekes, L. (2012). Integral bridge: A review on its behaviour under earthquake loads. In F. Biondini & D.M. Frangopol (Eds.), *Bridge maintenance, safety, management, resilience and sustainability.* London: CRC Press/ Balkema, Taylor and Francis Group, Stresa, Italy, July 8–12.

Soubry, M.A. (2001). *Bridge detailing guide.* London: Ciria.

Springman, S.M., & Norrish, A.R.M. (1996). *Integral bridges. Researchers' viewpoint.* Seminar on the design of integral bridges. London: Institution of Structural Engineers, jointly with IABSE and Highway Agency.

Taylor, H.P.J. (1993). Continuity in decks with precast beams – Practical issues. *Proceedings of the Henderson Colloquium "Towards Joint Free Bridges",* Pembroke College, Cambridge, UK, 20–21 July 1993, B. Pritchard (Ed.), E & FN Spon, Chapman and Hall, London.

Westergaard, H.M. (1948). New formulas for stresses in concrete pavements of airfields. *ASCE Transactions, 113,* 425–444, Paper No. 2340.

Yettram, A.L., & Husain, H.M. (1965). Grid-framework method for plates in flexure. *Journal of the Engineering Mechanics Division, 91,* 243–251.

Pathology, appraisal, repair and management of old prestressed concrete beam and slab bridges

Bruno Godart

The beam and slab bridges named in France VIPP are viaducts with multiple single spans made of prestressed concrete beams that are precast on site and post-tensioned, and then assembled transversely by prestress. A great number of those bridges were built in France at the beginning of the development of the prestressing technique. Some of them present strong losses of prestress related to the corrosion or the rupture of tendons. After a presentation of the pathology and its various causes, the paper presents the difficulties of the appraisal and the methodology that has been developed to assess their residual load-carrying capacity. The appraisal method is based on several levels of investigations and re-calculations. Then the paper addresses the problem encountered with the management of such structures by emphasising the interest of the risk analysis for helping owners to optimise the management of such bridges, with respect to safety. Finally, different solutions of repair that have been used with success to maintain or to strengthen the capacity of these structures are reviewed.

1. The history of prestress in France

The oldest concrete structure prestressed by post-tensioning in the world is the experimental arch of the Veurdre bridge on the Allier river in central France. This arch, built by Eugene Freyssinet in 1908 and preserved until our days in the private property of an inhabitant of Moulins, has a span of 50 m, a height of 2 m and is supported on abutments linked by a tie prestressed by means of several hundreds of post-tensioned wires which introduce a compressive force ranging between 25 and 30 MN. Openings of windows operated in 1993 made it possible to note the good conservation of the wires which were simply laid out in grooves filled with sand and sealed by a hydraulic mortar. It shall be noticed that the whole tie had remained buried in the ground during several decades without any waterproofing on the surface of the tie (Figure 1).

If Eugène Freyssinet built many concrete arch bridges before 1928, it is, however, this date which marks the official beginnings of the prestress in France with the patent taken out by Freyssinet on 2 October and 19 November 1928. But it is only after 1939 that the real industrial beginnings of prestress may be witnessed with the supply of high-strength steel by the steel industry, and with the development of two innovative devices by Freyssinet: the anchoring of wires by conical friction and the prestressing jack with double effect. Unfortunately, because of the World War II, the development of prestress was stopped in France, and one may only notice the construction of three bridges prestressed by post-tension according to the Freyssinet process: the two slab bridges of Elbeuf sur Andelle (1942) and Longroy (1943), and the famous Luzancy bridge (1941–1946) (Figure 2) on the Marne river (a portal leg bridge with a 54 m span built using small prefabricated segments assembled by filled-in joints and three-dimensional internal prestressing), where the prestress of this bridge appears in good condition and where the thrust at the base of the legs was adjusted in 1997 to cope with the displacement of one abutment.

This period 1930–1950 was 'dominated' by Freyssinet who took as much care as possible to design a structure and to choose the materials, and who took care personally of the great quality of the construction of his bridges. He was thus aware that the prestressed concrete was a demanding technique as regards quality, and the evolutions which followed this period were going to point it out.

It is only at the end of the World War II that the large prestressed concrete bridges developed, in a non-regulated context, the euphoria of the reconstruction, and during a period marked by cement and steel shortage. The first builders oriented themselves towards the prefabrication of beams, and this gave place to the construction of simply supported spans, and then to viaducts with simply supported spans made up of beams prestressed by post-tension.

Figure 1. The experimental arch of the Veurdre bridge with its prestressed tie in the bottom.

2. Description and pathology of post-tensioned beam and slab bridges

A great number of beam and slab bridges were built in France after the end of the World War II (about 250 between 1945 and 1957, and 450 between 1957 and 1967). Most of these structures (including the two longest: the Saint-Waast bridge at Valenciennes, with a span of 64 m, and the Hippodrome bridge in Lille, with a span of 66 m) were built after 1947 in the absence of formal design rules. It was not until 1953 that the first circular on prestressed

Figure 2. The Luzancy bridge over the Marne.

Figure 3. Example of a beam and slab bridge (VIPP bridge), the Merlebach bridge.

concrete appeared, and not until 1965 that the Temporary Instruction No. 1 (MTPT, 1965) was published.

2.1 Description of beam and slab bridges

The beam and slab bridges (referred in France according to their acronym 'VIPP': Viaducs à travées Indépendantes à Poutres Précontraintes par post-tension) are viaducts with multiple single spans made of post-tensioned concrete beams (SETRA 1967). These bridges are made of simply supported spans, and each span is composed of precast post-tensioned concrete beams (generally precast on site) that are linked together by crossbeams and a slab on the top. According to the design, the crossbeams and/or the slab may be prestressed transversally (SETRA 1996). The length of the spans is generally ranging between 30 and 50 m (Figure 3).

These bridges are slim and piecemeal structures that often suffered from a poor grouting of their prestressing sheaths, from defects of sealing behind tendon anchorages and from a lack of or failing waterproofing membrane. However, very often, the concrete of these decks was of good quality. We are now detailing the pathology of these VIPP according to their three possible causes of defects: design, construction and maintenance.

2.2 Pathology of beam and slab bridges

2.2.1 Design defects

The enthusiasm which prevailed when the first generation of prestressed structures was being built was reflected in a total confidence in full prestressing (i.e. lack of cracking) and in a belief that the concrete under compression was watertight. Hence, a number of details appeared such as (HA, SETRA, TRL, & LCPC, 1999):

- lack of waterproofing (deck waterproofing only became mandatory after 1966);

Figure 4. Condition of a transversal tendon located in the slab below the drainage channel.

- lack of sealing behind tendon anchorages;
- lack of provision for drainage (these are often beams located under drainage channels which are most affected by corrosion (Figure 4));
- leaking expansion joints (Figure 5);
- piecemeal construction resulting in a large number of unprepared construction joints which could give rise to cracking due to restrained shrinkage;

- use of unprotected transverse tendons in grooves in the deck;
- use of sheaths made from bitumen-coated Kraft paper wrapped around the tendons which made grouting impossible (metal sheaths were not used until the late 1950s);
- large numbers of tendons per span in older structures with deck anchorages (e.g. typically up to 15 out a total of 20–25) increasing the number of points of possible water ingress;
- use of lead-lined steel ducts between 1950 and 1975 to reduce friction sometimes causing tendon corrosion as a result of bimetallic action between the ducts and the steel;
- use of prestressing steel between 1950 and 1970 which was susceptible to stress corrosion, which gave rise to the possibility of brittle fracture;
- insufficient concrete cover to the reinforcement, resulting in corrosion of the reinforcement and spalling of the concrete giving easier access to the prestressing tendons for aggressive agents (Figure 6).

Moreover, cracking may frequently follow the longitudinal line of tendons in the webs and flanges of

Figure 5. Disorders due to leaking expansion joints.

Figure 6. Honeycombs and concreting defects at the soffit of a beam showing prestressing sheaths.

the beams. These cracks begin near the anchorages, propagate when the tendons are tensioned and then stabilise over time. Although they do not give rise to any special risk from a structural point of view, they can provide pathways for the movement of water within the interior of the structure. Calcareous deposits are frequently observed along some of these cracks (Figure 7). They can also provide entry points for water when poor drainage arrangements allow it to be directed to the outer surface of the edge beams.

Design defects are therefore mainly linked to construction defects or unsuitable techniques, or to the use of low durability materials. But the construction and maintenance defects have also a great part in the development of the corrosion of tendons, as the two following paragraphs derived from (HA, SETRA, TRL, & LCPC, 1999) are going to show it.

2.2.2 Construction defects

Construction defects which can cause corrosion of the prestressing tendons are unfortunately encountered rather frequently. The two main defects leading to corrosion are poor waterproofing (Figure 8) and incomplete grouting of the ducts (Figure 9). It is common to find thin waterproofing layers which do not extend beneath the footways. It is also common to find partly empty or sand-filled ducts in structures built before 1960 when grouts used to contain sand. Grouting techniques have also been poorly applied, and evidence of blockages has frequently been found. In addition to these defects, poor sealing of beam end anchorages, deck anchorages and transverse anchorages are observed.

Although concrete has generally performed well in older structures, there may be areas where poor workmanship has given rise to honeycombing or shrinkage cracking. The most common locations are the flange soffits where concreting has been made difficult by the

Figure 8. Damages at the bottom flanges of VIPP beams due to a defective waterproofing membrane and the presence of leakage through blasting devices made of steel tubes inserted in the structure.

congestion of ducts, or where there has not been proper compaction, resulting in large areas where spalling of concrete may allow aggressive agents to penetrate.

2.2.3 Defects linked to maintenance and operation

During the 30 years which followed the end of World War II, the absence of an inspection policy for structures and the failure to take the necessary actions to make the structures watertight are the two main causes of maintenance problems. Resurfacing the carriageway without investigating the condition of the waterproofing layer, failure to restore drainage to its required condition and clearing obstructed drainage pipes, and failure to act to prevent water leaking through expansion joints and running over the ends of beams are some of the many maintenance defects which have encouraged the development of corrosion in the tendons. In addition, the increasing use of deicing salts on the carriageway surface

Figure 7. Crack following the tendon line.

Figure 9. Incomplete grouting of a duct.

exacerbated the aggressive nature of the water seeping into the structure.

2.3 *The corrosion of prestress*

The corrosion of prestress can be divided into two main types:

- electrolytic corrosion (rust, small pits, generalised corrosion, etc.);
- stress corrosion cracking (including hydrogen embrittlement corrosion).

Most corrosion defects are linked to the first type of corrosion (conventional or general corrosion) and are caused by water which seeps through cracks and which infiltrates into the structure through leaking seals or zones of porous concrete. Then water flows through the network of ducts which have been grouted to a greater or lesser extent, etc. Experience shows that problems due to corrosion are distributed randomly, in accordance with the random nature of the movement of water within a structure. Figure 10 taken under the soffit of a beam after removal of the concrete cover is a perfect illustration of this 'random' distribution; we can observe that healthy tendons

Figure 11.   Brittle failure of wires of the KA prestressing kit.

are close to heavily corroded tendons in the same cross section, and that healthy zones alternate with corroded zones on a same tendon. The corrosion of tendons is significantly exacerbated by the presence of chlorides in the flowing water, but in most of the studied cases in France, relatively few or even no chlorides have been found in steel corrosion products taken from tendons on site.

Brittle fractures attributed to stress corrosion cracking have been found in some cases. Investigations conducted by Laboratoire Central des Ponts et Chaussées (LCPC) show that stress corrosion cracking appears primarily in quenched hot rolled wires, and more often in such wires containing more than 0.1% of Cu. This type of prestressing steel was only used from 1950 to 1965. We are especially cautious with some type of wires named 'sigma-oval' used with the KA prestressing process (Figures 11 and 12). Stress corrosion cracking due to hydrogen embrittlement has never been found in any of the surveys on French bridges. Likewise, no fatigue fractures of tendons or wires have so far been discovered in existing prestressed concrete structures.

Although defective grouting of the ducts is a primary condition for the development of corrosion, this condition on

Figure 10.   Illustration of the 'random' distribution of corrosion due to the random circulation of water inside the ducts.

Figure 12.   Brittle failure of a 'sigma-oval' type wire.

its own is not sufficient to trigger the reaction. It has been found that ungrouted tendons in dry structures located in regions with a continental climate have not corroded at all. Experience shows that neither the steel sheath nor even a well-compacted grout can form a sufficiently tight barrier against the percolation of an aggressive water through the more or less porous concrete of a beam. This was particularly well highlighted during the expertise of a viaduct made up of simply supported box girders of good quality, located in a port, whose bottom of the box girders was immersed in sea water only when very strong tides occurred, and where the corrosion of tendons at mid-span was such that the bridge had to be rebuilt. This was also observed on another viaduct whose prestressed piers bathed permanently in sea water.

Although the evidence of tendon corrosion arising from a deck waterproofing defect usually shows up as simple water leakage or the presence of seepage products such as efflorescence or traces of rust, there are a number of cases where tendons have been partly corroded without any external signs or visible defects. Finally, the risk created by corrosion of prestressing tendons is greatest in twin-beam bridges without intermediate cross-beams. The risk is even higher in the areas subject to high shear forces where brittle fracture is likely because there is insufficient passive reinforcement to carry the tensile forces transferred to the steel as the concrete cracks.

In addition to the question of the physico-chemical condition of tendons conservation, another significant question is to know the mechanical state of conservation of these tendons, and in particular to know if the tension in a healthy wire could be maintained during time. A partial response to this interrogation is provided to us by B. Tonnoir of the Laboratoire Régional des Ponts et Chaussées in Lille: since the 1980s, he carried out many tension measurements on VIPP tendons using his cross-bow method, and he assessed that the current residual tension of wire of diameter 5 and 7 mm used in the 1950s remains inside an interval ranging from 80 to 90 kg/mm^2, for an initial tension of the wires equal to 110 or 115 kg/mm^2 and a final tension estimated at the time of construction to be equal to 85 kg/mm^2 (Tonnoir, 2010).

3. The methodology of diagnosis

Some of these VIPP present strong deficits of prestress related to the corrosion or the failure of tendons, not detected during traditional examinations, because no external sign makes it possible to reveal them. It is particularly the case of the prestressing steels sensitive to stress corrosion cracking or to hydrogen embrittlement, and more particularly the case of the KA post-tensioning kit (sigma oval wires).

Moreover, in the first generation of VIPP, longitudinal reinforcement steels were much reduced (as well in area as in number), and vertical reinforcement steels close to the

supports were insufficient. This weakness of reinforcement steels involves a lack of ductility of the structure during the failure of prestressing tendons, and in this case the ruin of this type of structure can then be fast and fragile (Godart 2003).

For these VIPP, the occurrence of shear cracks or bending cracks corresponds to such a reduction in the prestressing force that the safety margin with respect to the ruin is low, and that very often the only solution is to condemn the bridge in the absence of reliable means of diagnosis, making it possible to assess the residual prestress and its evolution. Moreover, knowing that there is not any non-destructive testing methods to assess the global state and the quantity of residual prestressing all along a span of VIPP, one can consider easily that the diagnosis methodology for the VIPP is difficult to conceive and to carry out. Nevertheless, we propose a methodology which is strongly inspired of the LCPC guide (LCPC, 2001) and improved by the knowledge gathered since 2001 (Cremona 2007). This methodology of diagnosis comprises the three following stages.

3.1 Stage 1: Analysis of the bridge records

The first stage of the diagnosis consists in identifying in the bridge file all the records that are useful for the diagnosis and in particular those which make it possible to carry out a preliminary analysis of the risk:

- construction year (a bridge built before 1967 is *a priori* doubtful);
- design assumptions and calculation assumptions;
- post-tensioning kit and type of unit (tendons, anchorages, ducts, etc.);
- grout composition;
- grouting method used and incidents of injection and
- waterproofing system.

3.2 Stage 2: Detailed inspection of the bridge

This second stage aims to continue the preliminary analysis of the risk and especially to assess the risks of defects of integrity of the tendons. It consists in detecting the presence of disorders which can make fear a corrosion or a failure of tendons, such as for example:

- transverse cracks starting from the bottom part of the beam and generally going up in the web vertically (flexure) and sometimes with a slope close to the supports (shearing force);
- longitudinal cracks following the tendons in the webs, with traces of water circulation in the cracks;
- longitudinal cracks following the tendons in the bottom flange with traces of water circulation in the cracks;
- detachment of sealings behind anchorages of longitudinal or transversal prestressing and so on.

However, as already mentioned, the problems of these bridges are that it is not because none of these disorders is

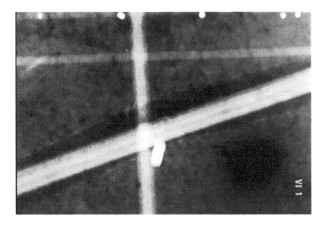

Figure 13. Deformed duct and lack of grout.

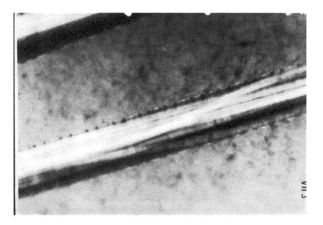

Figure 14. Gammagraphy showing a lack of grout and wire failures on a tendon in a corrugated duct.

apparent that prestressing is healthy. Regardless of the result of the detailed inspection, it is thus appropriate to go to the following stage which comprises three levels of investigations.

3.3 Stage 3: Investigations

The investigations mainly based on non-destructive techniques, but also on partially destructive techniques (Godart and Chatelain 1997) are graded according to three levels N1–N3.

3.3.1 Level of investigation N1

This level allows to evaluate the quality of grouting of the prestressing ducts. This evaluation is done by gamma-graphy or radiography examinations which make it possible to detect:

- the presence or the lack of grout (Figure 13);
- the conformity of the layout of the ducts (locations, crushed ducts, abnormal curvatures);
- the possible presence of wire or strands failures (non-systematic detection and so on) (Figure 14);
- the conformity of the reinforcement.

Nonetheless, these techniques cannot detect the presence of corroded wires.

It should be noted that the radar (ground penetrating radar applied to investigations on structures) also makes it possible to assess the conformity of the layout of the ducts and reinforcement, but it does not allow in any way to assess the presence or the absence of grout in metal sheaths. It can make it possible to position the gamma or radiographic pictures. It may also be noted that the use of the impact-echo method allows the detection of large grouting defects in metal sheaths, without being able, in the current state of the technique, to detect small voids. According to the importance of the lack of grout detected for a given

percentage of ducts, the LCPC guide (LCPC, 2001) leads to either a normal surveillance or to pursue to level N2.

3.3.2 Level of investigation N2

This level allows a qualitative evaluation of the prestress by opening access ports in order to inspect the state of the tendons. In particular, it allows an examination of the type of corrosion: simple rust, pits of corrosion, craters, generalised corrosion and stress corrosion. It also makes it possible to take samples of grout, water, sheath, wires, strands, rust, etc. for the purpose of laboratory analysis.

The opening of windows also allows a practice of the test known as 'of the flat screwdriver' (Figure 15). This test consists of trying to introduce the extremity of a flat screwdriver between the wires of a prestressing tendon and exercising a torque on the screwdriver handle. If the screwdriver blade rotates, it means that the wire is detensioned or even broken beyond the area where the window was opened. If the results are satisfactory (absence of corrosion, lack of failures and water in the sheaths), the LCPC guide (LCPC, 2001) advises to proceed with a normal surveillance.

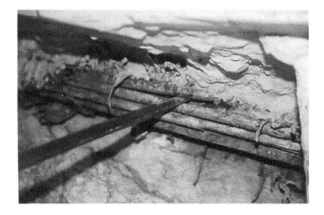

Figure 15. Test of the 'flat screwdriver' on site.

Figure 16. Measurement of the tension of a prestressing wire with the crossbow method.

Figure 17. Stress release method: view of the saw with a flat jack in the slot.

If the results show stress corrosion, or a generalised corrosion on at least 5–10% of the prestress, or if the analysis of the materials shows risks of corrosion, then it recommends to pursue to level N3.

3.3.3 Level of investigation N3

This level aims at assessing the residual load-carrying capacity of the bridge by practising various testing methods of the prestress and the structure, in order to allow a recalculation of the bridge. The various testing methods that can be used are the following:

- The measurement of tension of the prestressing steels by the crossbow method (Tonnoir, 2009). Its principle is to deviate a wire or a strand from its route and measure the force required for its deviation; this force is theoretically proportional to the tension existing in the wire or strand. This method is particularly rich in terms of knowledge acquisition because it allows to assess the residual tension in wires or strands, and this whatever the degree of corrosion affecting the prestressing steel (Figure 16). It also provides an invaluable average tension for the recalculation.
- The evaluation of the local longitudinal stresses by the stress release method: this method allows to estimate the stress gradient existing in the thickness of an element at a given point (Figures 17 and 18). It is possible to obtain the normal force existing in a cross section of a VIPP beam, with the condition of carrying out a minimum of three to four slots; Figure 19 presents an example of the implementation of three slots located in the upper flange, at mid-height of the web and above the lower flange, where the mean stresses measured are, respectively, 7.3, 6.5 and 9.8 MPa. This example shows not only that the profile of existing stresses in a non-loaded beam of a VIPP is nonlinear, but also that the

stresses are increased in the thicker parts of the cross section and lowered in the thinner parts (Figure 20). The explanation for this behaviour lies in the redistribution of stresses within the section due to the delayed deformations induced by the creep and shrinkage of concrete. Figure 20 shows the differences that may exist between the stresses estimated by the necessarily simplistic calculations made by most of the consultants and the stresses existing really in the structures. It also shows the limits of the method for assessing global efforts in a cross section of a beam, because several measurement points are required in order to make a proper integration of stresses on the whole cross section (Abdunur, 1993).

- The curvaturemetry: this method allows to detect non-linearities of curvatures under bending moment indicating the beginning of a not yet visible cracking in the central zone of a beam. It is particularly useful to detect sections which would be close to decompression. For the method to be effective, it is, however, necessary to be able to load the bridge

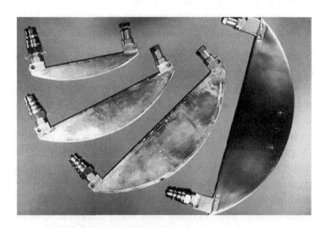

Figure 18. Various flat jacks for inserting in the slots to obtain the stress gradient in concrete.

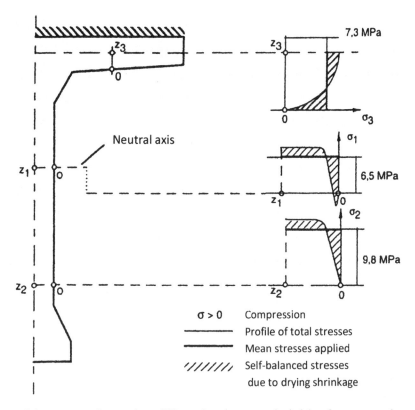

Figure 19.   Measurement of the stress gradient at three different locations over the height of a prestressed concrete beam.

heavily using trucks, without being certain to achieve the targeted objective, but at least one knows that under such a loading there was no abnormal behaviour of the curvature. The cracked zone in the central part of the beam being likely to extend over a rather big length, it is often necessary to place several curvaturemeters in a row simultaneously (Figure 21).

The acoustic monitoring can also be used to survey a VIPP and to listen to the elementary wires or strands failures.

If it does not provide directly useable results to assess a load carrying capacity, it, however, allows to confirm the existence of a prestress damage by corrosion and to follow it with time. The final evaluation is done on the basis of recalculation integrating the results of the investigations and in particular the estimates of prestress losses. This recalculation can be done by traditional methods and if necessary on the basis of a reliability approach which requires to know the probabilistic distribution laws of certain calculation parameters (Cremona 1995).

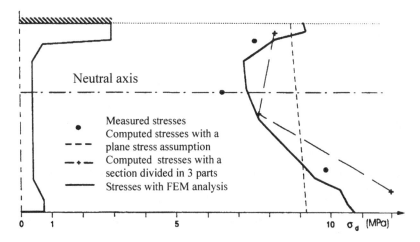

Figure 20.   Comparison of stresses measured by the stress relieve method with stresses computed by different methods.

Figure 21. Installation of several curvaturemeters on the deck of a bridge.

Figure 22. Injection of a corrosion inhibitor in a duct by powerful ultrasounds (Courtesy ATEAV Company).

According to the results of the recalculation and investigations of level N1, N2 and N3, one can proceed with various possible types of treatment:

- a reinforcement by steel plates, by composites or by additional bars embedded in shotcrete, particularly for insufficiencies of shear capacity;
- a strengthening by additional prestressing for insufficient resistance in bending and in shear close to the supports;
- a reconstruction of the most damaged spans or beams;
- an injection of grout in the empty ducts when this is possible (technique of the vacuum injection), or the injection of a corrosion inhibitor by powerful ultrasounds (Figure 22) (Dubois & Michaux, 2008);
- a repair of the waterproofing membrane, drainage, sealings behind anchorages, etc.

4. Risk analysis applied to the management of VIPP

4.1 Introduction

Risk analysis is a powerful tool to achieve certain strategic choices when financial resources of the authority are not expandable. It is particularly the case when the authority has to manage a great quantity of old VIPP that is difficult to appraise as discussed previously, and the authority has to do priorities to investigate among the bridges. In the risk analysis methods, several risk categories are identified:

- the risk of collapse of whole or part of the structure: it is the largest and most serious risk that has human, economic and environmental consequences;
- the risk of loss of serviceability of the infrastructure: the VIPP supporting a motorway does not collapse but it cannot be used for a certain duration, which has important consequences on the socio-economic activity of a region;
- the financial risk on the total cost of the project, on its expected life due to construction hazards or on an inability to future needs, etc.;

- the risk of evolution of the external environment (e.g. new use of an existing motorway as a city ring road).

In France, the first application of risk analysis on bridges was conducted in 2007 by C. Cremona at LCPC on the VIPP of the conceded motorway network on behalf of the French Association of Motorway Companies (ASFA) (Dabert & Cremona, 2009; Guérard, 2009). This association decided in 2005 to launch a bid to study the reliability of the knowledge of their 116 VIPP. LCPC (now IFSTTAR) then proposed to apply a standard approach founded on operational safety (or dependability). The operational safety analysis is based on a qualitative and quantitative engineering activity. Originally designed for the phases until the production stage, particularly in the field of nuclear engineering, it focuses on risk identification, qualification and simplified quantification to formulate proposals for detailed quantification of risks and corrective solutions.

4.2 The operational safety analysis of the ASFA VIPP

4.2.1 Stages of the operational safety

The operational safety analysis proposed by LCPC has been divided into four stages:
Stage 1: an internal functional analysis (IFA) which covered the following aspects:

- the detailed inventory of the concerned structures, including their technical characteristics;
- (date of construction, company, engineering and design, dimensions, waterproofing, etc.);
- the recovery of data files on bridges (stocktaking and monitoring, maintenance and repairs, detailed inspection reports, dates and principal conclusions,

list of works, traffic supported) and the inventory of the possible modes of exceptional operation modes of each bridge;

- the identification and analysis of repairs already carried out (attachment of spans, change of bearings, etc.);
- the analysis of the bridge environment (aggressiveness, crossing, access, network environment, winter viability, frequency and nature of deicing products, etc.);
- the analysis of constituent materials (concrete, prestressing, grouting, duct, etc.) and methods of construction;
- the analysis of ancient calculation methods (regulations, trafficable width, overload);
- the inventory and analysis of investigations conducted.

Stage 2: a preliminary risk analysis (PRA) with the objective to:

- define families of structures based on relevant criteria (construction, failure, etc.);
- identify gaps and problems specific to each bridge and each family.

Stage 3: a risk quantification (RQ) aiming at:

- checking by simple calculation, the ability of the current configuration of bridges to support the operational modes;
- defining the possible interim operating modes of the failed deck as a function of the probable visible defects (no operation, one line or two lines for heavy weight vehicle and/or light weight vehicle).

Stage 4: a detailed analysis of criticality to:

- identify and propose further investigations that are necessary to characterise the condition of each bridge and each family;
- define a method for assessing the residual load-carrying capacity and the remaining life of the bridge according to different operating modes, in order to be able to decide whether the whole traffic can be maintained or traffic loads have to be restricted.

4.2.2 Methodology and main results (Dabert & Cremona, 2009)

4.2.2.1 Internal functional analysis. The data inventory was conducted and organised on the basis of a data file for each bridge which allows the identification of all data necessary to properly conduct the further analysis. The files are computerised in a format such as Excel© to facilitate the multi-criteria analysis of the PRA. Across the network, eight bridges have also been the subject of an anomaly file, because they were inconsistent with the VIPP typology.

4.2.2.2 Preliminary risk analysis. The PRA is a deductive analysis based on information available in the bridge records that allows to have visibility on the shortcomings of bridges and provides a hierarchy of bridges in five classes. It allows to concentrate the analysis and detailed investigations on the most sensitive structures. It seeks both to identify the factors or events that could reduce the bridge performance, and to affect to bridges a grade of severity according to predetermined criteria.

To rank the risks in the VIPP, it was proposed to consider that a structure built with the current load and calculation regulations, with a modern design and an overall non-prestressed slab, presents no special risks, except design and execution defect inherent to any bridge. A design involving no particular risk receives a grade of zero. From this reference state, any factor is generally assessed on a scale with four levels of criteria ranging from $+1$ to $+4$. Some factors have a larger grading, as is the case of the factors 'Condition of the prestress' and 'IQOA rating' (Robichon, Binet, & Godart, 1995) that have more extensive grading. Finally, improvements done to the structures can improve their performance, which led to retain negative gradings.

The preliminary risk analysis highlights the sensitive classical themes like the overall design, the construction, the operation and maintenance, and the environment. It should be noted that the strength of structures is determined directly by the prestressing of the structure, which justifies a theme in itself.

(a) *The general design*

The general design of the first generation of VIPP is much less satisfactory than the current design because of longitudinal concrete joints along the beams, low passive reinforcement, too thin webs and lack of small slabs of continuity on piers. The date of construction is by itself an important criterion (first generation, tendons susceptible to stress corrosion, etc.). As part of the PRA, the following qualitative criteria were selected to characterise the overall design of a bridge:

- bridge with overall deck slab or intermediate slabs;
- bridge with or without transverse prestressing;
- presence or absence of a substantial shear reinforcement in beams;
- presence or absence of a substantial passive reinforcement in the deck slab;
- thickness of the web in the cross section near the support at mid-height;
- thickness of the web in current section;
- presence of small slabs of continuity, which help prevent water linkage on the intermediate supports, with or without intermediate crossbeams;
- number of beams (if the bridge has few beams, the low redundancy of beams increases, for the

structure, the severity of the consequences of disorders occurring on a beam);

- quality of water drainage;
- prestressed hammer-type piers and
- absence and/or type of waterproofing.

(b) *The prestress*

The prestress has been the object of two sub-themes of specific analysis. Corrosion of prestressing tendons is indeed the primary concern on the VIPP, as with all prestressed bridges. We must be able to characterise the proper protection of tendons (sealing, grouting) and the risk of stress corrosion cracking. The presence of inclined tendons that are anchored in the deck slab is a detrimental feature to the durability of VIPP. The age of the prestressing governs the risk because of the nature of the steel used, duct, grouting products and regulations.

(b1) The sub-theme 'initial prestress' takes into account the evolution of the regulations, both on the traffic loads and on material stresses because they affect the amount of prestress implemented and thus the real safety factor of the structure. That is why the following criteria were used:

- prestress calculated before the IP1 rule (1965), between IP1 rule and BPEL 1983 code, and after BPEL 1983 code;
- bridge built before the load regulations of 1960, between 1960 and 1971, or after 1971 (for the local bending).

(b2) The sub-theme 'prestress condition' is characterised by the following criteria:

- apparent condition of the prestress (not corroded, somewhat corroded, fairly corroded, very corroded) assessed on the most degraded area of the structure;
- grouting of ducts (filling condition of ducts);
- apparent condition of the concrete cover;
- management of the waterproofing during the life of the structure;
- presence or absence of a small slab of continuity between two spans and
- use or not of deicing salts.

(c) *The maintenance and the environment*

They are also characterised by criteria that the length of this article does not allow to detail (see Dabert & Cremona, 2009).

(d) *The general note*

It is obtained from the severity notes identified in the preceding paragraphs, with a weighting system. Following tests on a dozen of bridges, a choice of weighting coefficients by theme was proposed and accepted, leading to classify bridges into five classes of risk (Table 1).

The bridges are mostly classified in the class of 'moderate' risk.

Table 1. Overall qualitative index of a bridge (PRA).

Classification of bridges	Note
No risk	0–20
Moderate risk	20–40
Rather high risk	40–60
High risk	60–80
Very high risk	80–100

4.2.2.3 Risk quantification. The objective is to assess the suitability of the current configuration and of deteriorated configurations of structures, and then to compare them within their families. To do this, the method must be sufficiently rich to describe the behaviour of the bridge, but, however, as simple as possible. Quantifying the mechanical criticality is therefore based on simple indicators to assess the serviceability (SLSs: service limit states) and the structural safety (ULSs: ultimate limit states). In this goal, this criticality is represented by a criticality indicator I of a generic type:

$$I = \frac{\text{Resistance}}{\text{Load}}.$$

This indicator is assessed for a bridge used in its initial configuration, in its present situation, then with reduced lanes. The evolution of the indicators is estimated for deteriorated situations (losses of prestressing tendons that are assumed or effective according to the data inventory) or for limited traffic in weight.

The methodology consists of approaching the real behaviour of the bridge, through a simplified but realistic assessment of the structure. The effects of loads that are fixed or variable, are approached as finely as possible from calculations by structural parts (beams, slabs, hammer-type piers), to ensure that the simplified model approaches the complex behaviour with a difference less than 10%.

Due to the complexity of a VIPP-type structure, a single indicator cannot summarise the behaviour of the bridge. Therefore, one should look for specific remarkable elements that together contribute to the structural safety of the structure. It is the resistance of the slab receiving the traffic loads, the resistance of the longitudinal main beams and then the resistance of the supports. For beams that are in a single span, we consider that a good evaluation of their resistance is to control their flexural strength at mid-span and their shear strength on support.

These indicators clearly cannot, by themselves, assess the load-carrying capacity of the actual structure that is only available (for a complex bridge built by phases and with prestress) by a full recalculation, beam after beam, and with the condition to provide realistic information on the tension of prestressing tendons. They are a tool available for a manager to assess roughly the intrinsic level of service of his bridges, but more importantly, to

Table 2. Classification of the bridge condition based on the percentage of prestressing (ULS).

Bridge condition	Level	Index I	Percentage of necessary prestress/initial prestress
Robust	1	$I > 1.25$	$p < 90\%$
Probably conform to a modern design	2	$1.15 < I < 1.25$	$90\% < p < 100\%$
Probably undersized, but able to sustain the loads	3	$1 < I < 1.15$	$100\% < p < 115\%$
Undersized and probably not able to sustain loads on a regulation basis	4	$0.85 < I < 1$	$115\% < p < 130\%$
Undersized and unable to sustain loads on a regulation basis	5	$I < 0.85$	$p > 130\%$

prioritise the serviceability level of structures and to establish a management policy for the maintenance of the patrimony.

Quantitative indicators make the comparison between resisting efforts and loads. The higher the indicator, the greater the bridge part presents a significant resistance with respect to the situation of service examined (type of load, regulation type and operating conditions). Evaluation of indicators comes from calculations comparing the actions depending only on the operating conditions of the structure, with resisting efforts that depend only on the intrinsic characteristic strength of constituent materials of the structure and their position in the structure. In the case of prestressed concrete structures, the resisting efforts are highly related to the prestress integrity, yet access to this basic data remains very difficult.

The proposed indicators relate to the bending of beams, the shearing force of beams, the bending of the deck slab and the bending of the hammer-type piers, with respect to SLS and ULS in each case. Tables 2 and 3 provide the classification of the bridge condition based on the percentage of prestress required in relation to the initial prestress. The overall index RQ is defined for the resistance of beams with respect to bending and shear, and for the resistance of the deck slab by retaining the maximum of the different coefficients calculated. A bridge is then ranked among the five classes defined in Table 4.

4.2.2.4 Analysis of criticality. The qualitative index reflects the overall susceptibility to potential or actual risks of the bridge and reports primarily for the assessment of the risk of prestress loss. The quantitative index reflects the overall mechanical health condition assuming that the structure is undamaged. For the manager, these indexes should therefore be crossed to try to assess the mechanical

condition given the likely condition of prestress. A matrix of decision-making has been proposed and is illustrated in Table 5.

From this table, it is possible to define overall criticality classes to outline a method of classifying bridges by assigning an overall and unique rating to each bridge. The overall criticality rating crosses information on the apparent condition of the structure (mainly characterised by the PRA classes) with information on the theoretical structure condition characterised mainly by the RQ note (Table 6). It is important to note that the classification results from a preliminary analysis, based on an analysis of archive records that may reflect only a partial reality (incomplete archive file, evolution of the structure insufficiently traced, relevant information not necessarily identified in gathering data, etc.) as well as an estimate of the RQ notes based on simplified assumptions and calculations.

The criticality analysis can then determine what actions will be undertaken on each of the VIPP according to its ranking. VIPP with criticality D will be treated in priority and subjected to heavy investigations and thorough recalculations. VIPP with criticality A will just be subjected to a normal surveillance. VIPP with intermediate criticality B or C will be subjected to recalculation or targeted investigations as defined in the LCPC guide on assessment of VIPP which proposes three levels of investigations N1 to N3 (LCPC, 2001).

4.3 The risk analysis applied to the bridges of the French State

Based on the success of this experiment, Sétra (Service d'Etudes des Transports et de leurs Aménagements) has proposed to the Road Directorate of the French State to

Table 3. Classification of the bridge condition based on the percentage of prestressing (SLS).

Bridge condition	Level	Index	Percentage of necessary prestress/initial prestress
Conform to a modern bridge and having a good durability	1	$I > 1$	$p < 100\%$
Probably undersized and can present disorders reducing the expected life	2	$0.85 < I < 1$	$100\% < p < 115\%$
Having a behaviour leading to a shorter life	3	$I < 0.85$	$p > 115\%$

Table 4. Overall quantitative index of the risk presented by a bridge.

RQ index	Class
1	Good
2	Fair
3	Correct
4	Poor
5	Bad

Table 5. Percentages of bridges in each pair (PRA, RQ).

PRA index	RQ index				
	1	2	3	4	5
0–20	0%	3%	4%	0%	0%
20–40	5%	8%	34%	7%	9%
40–60	3%	3%	14%	3%	5%
60–80	0%	0%	1%	0%	2%
>80	0%	0%	0%	0%	1%

develop a methodology for analysing risks for the specific types of bridges described as 'sensitive'; four families of structures were identified as priority handling:

- viaducts made of single spans with prestressed concrete beams (VIPP);
- steel culverts;
- earth reinforced structures and
- structures located in the aquatic site with a risk of scour.

The methodology of Sétra (Hyppolite & Billon, 2009; Sétra, 2010) uses three factors that are hazard, vulnerability and severity of consequences:

- The hazard is the phenomenon causing the risk. It may be environmental in nature (earthquakes), human (impact) or internal (corrosion). A hazard is characterised by its probability of occurrence and intensity; for the VIPP, the degradation of the prestress corresponds to a random internal damage to the materials. Probability and intensity will be qualified by a class of hazard.
- The vulnerability is the sensitivity of a structure to the studied hazard: for the VIPP, it depends on the initial design and the amount of damage. It will be quantified by a ratio between the structural

Table 6. Global criticality of VIPP.

	Overall criticality
A	Healthy bridge
B	Little deteriorated bridge
C	Deteriorated bridge
D	Very deteriorated bridge

resistance and the regulatory loads. This ratio is an indicator of robustness (the opposite of vulnerability). The study examines various scenarios of operating the structure by reducing the number of lanes and by limiting the traffic.

- The severity is generally gauged according to the human, economical and environmental consequences; for the VIPP, this indicator is primarily linked to the socio-economic value of the structure and to the roads that are supported or crossed. The owner will thus have a hierarchy of his patrimony and he will be able to consider, primarily, the bridges having a high probability of failure and a great strategic interest.

4.4 Conclusions on the risks presented by the VIPP-type bridges

An investigation by Trouillet (2000) shows that, over a total of 720 VIPP built before 1966, about 15 were already demolished because of corrosion of their tendons. Very fortunately, until now, we did not have to deplore any sudden collapse of VIPP spans, thanks in particular to a policy of monitoring and inspection which made it possible to detect sufficiently early the bridges presenting some risks. Nevertheless, the simply supported spans prestressed by post-tension are among the structures most difficult to monitor. Indeed, VIPP, and especially those of the first generation, are structures which resist thanks to the integrity of their prestressing, because they have only a small percentage of passive steel. Two principal types of rupture are to be feared: a rupture by bending and a rupture by shearing force.

Regarding bending, the first cracks appear when the deficit of prestressing is already very significant. If we consider only one beam, it is necessary to have lost 40–50% of the longitudinal prestressing force to see a bending crack appearing under the current live loads. When these live loads are removed, the residual prestress closes again the crack and this latter becomes almost undetectable with the eye. Thus, one cannot assume that the absence of bending cracks implies that the bridge is healthy.

If only one beam is strongly damaged, and if the transverse prestressing is in good condition, the other beams must be able to take again the loads supported by the failing beam. If all the other beams are strongly damaged, then the rupture of a beam involves the total ruin of the bridge. Between these two extreme cases, a multitude of intermediate cases may exist which are a function of the degree of conservation of the prestress of the slab, of the crossbeams and of each beam. But generally, in these cases, it seems that a mechanism of ductile ruin corresponding to a progressive damage of the structure may develop.

With respect to shearing action, the situation is notably different, because, in the zones close to beam ends, the risks of brittle fracture can exist. Indeed, under the current cases of loading, the resistance of the cross sections to the shearing force in these zones can be ensured, thanks to only the concrete strength, and without the help of the prestressing tendons which can be strongly corroded. Then, it is sufficient that an exceptional loading case occurs so that a sudden failure of the beam cross section occurs, if the residual section of the tendons is insufficient to support the efforts released during the cracking of the concrete.

5. The repair of VIPP

Several solutions of repair or strengthening may be applied to the VIPP according to their condition and their residual load-carrying capacity. The main solutions are summarised further in the text, based on a simplified qualitative approach of the corrosion condition of tendons, for the sake of the present article (in reality, a quantitative assessment of the corrosion is performed with the help of the investigation techniques presented previously).

If the corrosion of tendons is negligible, and that some spalling of concrete has happened due to the corrosion of

Figure 24. Gluing of a CFRP strip under the soffit of a VIPP beam.

the reinforcement, then a patch repair is sufficient provided that it has been checked that the carbonation of concrete and/or the ingress of chlorides have not penetrated beyond the rebars. Sprayed concrete can also be used in the case where a large amount of concrete spalling has occurred. In the case where the penetration depth of aggressive agents is beyond the rebars, the problem becomes difficult because the removal of the contaminated concrete could be impossible in some regions of the beam like the toe in the middle of the beams or the shear resisting zone near the ends of the beam and the bearings; electrochemical treatment may be applied, but with extra care due to the possible damaging of the prestressing steels. In the case of cathodic protection, the standard NF EN ISO 12 696 (AFNOR, 2012) gives some rules to avoid induced hydrogen embrittlement of the prestressing steels due to the low cathodic potentials imposed to the steel tendons. Concerning the other electrochemical treatments that are chloride extraction and

Figure 23. Gluing of a CFRP tissue on the web of a VIPP beam.

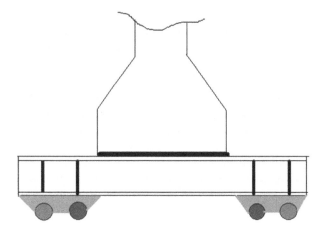

Figure 25. Cross section of the deviator in the middle of each beam.

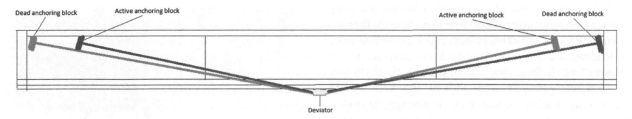

Figure 26. Installation of additional tendons with their dead and active anchoring blocks.

realkalinisation, we must be very cautious, particularly with chloride extraction, because the risk of hydrogen embrittlement is greater than for cathodic protection.

If the corrosion of tendons is small, or if the reinforcing steels are insufficient because of initial underestimation or loss of cross section due to corrosion, two repair solutions may be used:

- The first one is the gluing of steel plates: it is a proven technique that allows to add a cross section of reinforcement which is strong enough. It should ensure the respect of the ULS, but the improvement of the SLS will be limited since it only affects the moments of live loads. This technique is subject to a good condition of the concrete surface (pull-off test). The implementation is quite delicate because steel plates are heavy and they need to be laid on under the bottom flange of the beam with a certain pressure, and the traffic should be interrupted during the gluing phase and during the prolonged period of the resin hardening to prevent the plates from dropping, etc.
- The second solution is the gluing of tissues (Figure 23) or strips (Figure 24) made of carbon fibre-reinforced plastics (CFRP): this technique is increasingly used. Compared to the strengthening of a slab, the problem may be more difficult to solve on beams because the width of the flange is only 65 cm and it requires gluing several layers of tissues or thick bands. As for the steel plates glued, this technique requires a good strength of the concrete surface. The implementation is a bit easier than for the steel plates because of a lower weight and a setting by masking. Traffic should also be interrupted during the gluing phase and for the time of hardening of the resin (at least 24 h), which can be lower than for the steel plates.

If the corrosion of tendons is important, then the only solution is to strengthen by additional prestressing. This latter can be straight or deviated. The deviation of the prestressing is generally more efficient because it allows a strengthening of the shear zones and has a better efficiency to add bending moment, but it necessitates to add a deviator in the middle of the beam (Figure 25).

As illustrated by the following example of the Merlebach bridge in the east of France, a current strengthening consists of adding two tendons symmetrically on each side of the beam. Each tendon comprising two strands of diameter 15.7 mm is able to replace one corroded existing tendon of cross section 600 mm^2 (12 wires of diameter 8 mm and post-tensioned at 133 kg/m^2). One of the main difficulties is to anchor the new tendons at their ends. The anchor blocks must be installed in a suitable zone and preferably in the upper corner of the beam. Because of the presence of a crossbeam at the end of the span, and because there is no place behind the crossbeam to install a prestressing jack, the anchor blocks have to be fixed before the crossbeam and at a certain distance of this latter in order to have enough space to install the prestressing jack. To improve the strengthening system and to diffuse the anchoring efforts in the web of the beam, it is advisable to alternate dead and active anchoring as shown in Figure 26. Each tendon goes from a dead anchoring block located in a corner to an active anchoring block located at 3.84 m from the crossbeam.

The blocks are preferably precast on site (Figure 27) and are fixed to the beam with the help of very short prestressing bars that must be carefully post-tensioned to avoid a rapid loss of prestress (Figure 28). The tendons are

Figure 27. Concreting of precast anchoring blocks.

Figure 28. A dead anchoring block fixed on the web of a beam by two prestressing bars.

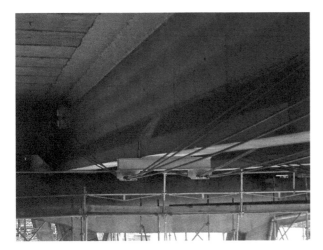

Figure 29. Installation of the additional deviated tendons.

tensioned as planned at one end only and there is no conflict between them because the two tendons are deviated in different planes at the deviator (see Figure 29).

6. Conclusions

To our knowledge, prestress is the only 'material' in civil engineering (and very likely in our environment) which works permanently to approximately 80% of its failure strength. It is therefore a sensitive 'material' which shall be protected as well as possible if one wants to have a certain durability. Pathology teaches us that the prestressing tendons are more vulnerable in the case of post-tensioning than in the case of pre-tensioning, and a better care should be granted to the grouting which is the key operation to ensure the durability of prestressing. But, it is particularly difficult to be ensured of the perfect filling of the ducts, even if non-destructive

methods like the gammagraphy is very useful to control the filling.

It was shown in this paper that the appraisal of old post-tensioned beam and slab bridges may be difficult, that the knowledge of the condition of conservation of all tendons with respect to corrosion is impossible to reach and that a risk analysis is helpful to assess and manage a patrimony composed with such bridges. It was also shown that these structures can also be repaired or strengthened with different techniques, and again prestress is one of the most valuable ones.

Lastly, a global analysis of the condition of bridges located on the French national road network shows that the condition of the prestressed concrete bridges is similar to that of the whole stock of bridges of the network, and that the durability of post-tensioning tendons is a problem which is classified in the average of the concerns of the managers. However, it remains that prestressing tendons are sensitive products which require detailed attention as regards quality of implementation, and this was already a conclusion of Freyssinet about a century ago.

References

Abdunur, C. (1993). *Testing and modelling to assess the capacity of prestressed bridge* (Vol. 67, pp. 353–360). IABSE colloquium on remaining structural capacity. IABSE report. Copenhagen. Zürich, Switzerland: International Association for Bridge and Structural Engineering (IABSE).

AFNOR (2012). *Protection cathodique de l'acier dans le béton* [Cathodic protection of steel in concrete]. Standard NF EN ISO 12696.

Cremona, C. (1995). *Mise à jour de la fiabilité des ponts précontraints au moyen de mesures de la force de précontrainte* (pp. 63–70). Bulletin de Liaison des LPC No. 199. Marne La Vallée, France: IFSTTAR.

Cremona, C. (2007). Suivi du comportement mécanique d'une poutre de VIPP sous chargement: Le cas du VIPP de Merlebach. Etudes et Recherches des Laboratoires des Ponts et Chaussées, Série OA No. 56. LCPC. Marne La Vallée: IFSTTAR. 154 pp.

Dabert, J.L., & Cremona, C. (2009, March 18–19). Analyse de la criticité du patrimoine VIPP autoroutier. In *GC'2009, Association Française de Génie Civil*, Cachan. Paris: Association Française de Génie Civil (French Association for Civil Engineering).

Dubois, P.M., & Michaux, D. (2008). *Technique innovante de traitement des câbles de précontrainte par injection d'un inhibiteur de corrosion au moyen d'une pompe à ultrason de puissance – Application aux viaducs à travées indépendantes à poutres précontraintes (VIPP) et à d'autres structures précontraintes*. Bulletin Ouvrages d'Art no. 59. Sourdun, France: CEREMA.

Godart, B. (2003). L'évaluation et la maintenance des viaducs à travées indépendantes à poutres précontraintes (VIPP). In *Application des notions de fiabilité à la gestion des ouvrages existants*, sous la direction de C. Cremona. Presses de l'Ecole Nationale des Ponts et Chaussées.

Godart, B., & Chatelain, J. (1997). L'auscultation des Ouvrages d'Art. In *Maintenance et réparation des ponts*, sous la direction de Calgaro J.A. & Lacroix R. Presses de l'Ecole

Nationale des Ponts et Chaussées. 666 pp. Paris: Presses de l'ENPC.

Guérard, H. (2009, March 18–19). Les VIPP du réseau ASF/Recalculs d'ouvrages: les VIPP d'ASF. In *Journées GC'2009, Association Française de Génie Civil*, Cachan. Paris: Association Française de Génie Civil (French Association for Civil Engineering).

HA, SETRA, TRL, & LCPC (1999). *Ponts en béton précontraint par post-tension* [Post-tensioned concrete bridges]. Joint publication of Highways Agency, Service d'Etudes Techniques des Routes et Autoroutes, Transport Research Laboratory and Laboratoire Central des Ponts et Chaussées (French and English), Thomas Telford. 164 pp.

Hyppolite, J.C., & Billon, J. (2009). Analyse des risques appliquée aux buses métalliques (pp. 26–33). Bulletin Ouvrages d'Art no. 61. Sétra.

Laboratoire Central des Ponts et Chaussées (2001). *Viaducs à travées indépendantes à poutres précontraintes (VIPP). Guide méthodologique de surveillance et d'auscultation*. Collective Guide, Marne La Vallée, France: IFSTTAR. 72 pp.

MTPT (1965). Circulaire No 44 du 12 Août 1965 portant approbation d'une instruction provisoire relative à l'emploi du béton précontraint dans les ouvrages dépendant du Ministère des Travaux Publics et des Transports (IP1). BOMEL, fascicule spécial No. 65-15 bis.

Robichon, Y., Binet, C., & Godart, B. (1995, August 23–25). 'IQOA': Evaluation of bridge condition for improved maintenance policy. In *IABSE symposium 'Extended the lifespan of structures'*, San Francisco (Vol. 73/1, pp. 407–412). Zürich, Switzerland: International Association for Bridge and Structural Engineering (IABSE).

SETRA (1967). *Dossier type VIPP*. Sourdun, France: CEREMA.

SETRA (1996). *Guide de conception des ponts à poutres préfabriquées précontrainte par post-tension (VIPP)*. Sourdun, France: CEREMA.

SETRA (2010). *Analyse des risques appliquée aux viaducs à travées indépendantes en poutres précontraintes (VIPP)*. Guide technique. Sourdun, France: CEREMA.

Tonnoir, B. (2009). *Mesure de la tension des armatures de précontrainte à l'aide de l'Arbalète*. Technical Guide. Marne La Vallée, France: IFSTTAR. 24 pp.

Tonnoir, B. (2010, October 19–20). Les mesures à l'arbalète: Méthodologie de l'essai, résultats. In *Colloque LE PONT, Association LE PONT*, Toulouse. Retrieved from www.le-pont.com

Trouillet, P. (2000). Communication interne, Mission d'Inspection Spécialisée Ouvrages d'Art. Secteur Autoroutes Concédées.

Deteriorating beam finite element for nonlinear analysis of concrete structures under corrosion

Fabio Biondini and Matteo Vergani

A three-dimensional reinforced concrete (RC) deteriorating beam finite element for nonlinear analysis of concrete structures under corrosion is presented in this study. The finite element formulation accounts for both material and geometrical nonlinearity. Damage modelling considers uniform and pitting corrosion and includes the reduction of cross-sectional area of corroded bars, the reduction of ductility of reinforcing steel and the deterioration of concrete strength due to splitting cracks, delamination and spalling of the concrete cover. The beam finite element is validated with reference to the results of experimental tests carried out on RC beams with corroded reinforcement. The application potentialities of the proposed formulation are shown through the finite element analysis of a statically indeterminate RC beam and a three-dimensional RC arch bridge under different damage scenarios and corrosion penetration levels. The results indicate that the design for durability of concrete structures exposed to corrosion needs to rely on structural analysis methods capable to account for the global effects of local damage phenomena on the overall system performance.

1. Introduction

The structural performance of concrete structures is time-dependent due to ageing and environmental damage (Ellingwood, 2005). Damaging factors include the effects of the diffusive attack from aggressive agents, such as chlorides, which may involve corrosion of steel reinforcement and deterioration of concrete (CEB, 1992). The direct and indirect costs associated with steel corrosion and related effects, in particular for concrete bridges and viaducts, are generally very high (ASCE, 2013; NCHRP, 2006). It is therefore of major importance to promote a life-cycle design of durable concrete structures and infrastructures (Frangopol, 2011; Frangopol & Ellingwood, 2010).

The experimental evidence shows that the main effect of corrosion is the reduction of the cross-sectional area of the reinforcing steel bars (Gonzalez, Andrade, Alonso, & Feliu, 1995; Zhang, Castel, & François, 2009). Corrosion can also significantly affect the ductility of steel bars and lead to brittle failure modes (Almusallam, 2001; Apostolopoulos & Papadakis, 2008). Moreover, the formation of corrosion products can involve deterioration of concrete due to propagation of splitting cracks (Vidal, Castel, & Francois, 2004; Zhang et al., 2009), delamination and spalling of the concrete cover (El Maaddawy & Soudki, 2007; Li, Zheng, Lawanwisut, & Melchers, 2007; Pantazopoulou & Papoulia 2001). In addition, the corrosion process may affect the steel–concrete bond

strength (Bhargava, Gosh, Mori, & Ramanujam, 2007; Lundgren, 2007).

The local effects of reinforcement corrosion can be limited based on simplified design criteria related to threshold values for concrete cover, water–cement ratio, amount and type of cement, among others (*fib*, 2006). However, a life-cycle design of durable structures cannot be based only on indirect evaluations of the effects of damage, but it needs proper methodologies to take into account the global effects of local damage phenomena on the overall system performance under uncertainty. To this aim, a probabilistic approach to lifetime assessment of concrete structures in aggressive environment has been proposed in previous works (Biondini, Bontempi, Frangopol, & Malerba, 2004a, 2006). This approach relies on general procedures and methods for time-variant nonlinear and limit analysis of concrete structures under static and seismic loadings (Biondini, Camnasio, & Palermo, 2014; Biondini & Frangopol, 2008).

Nonlinear analysis of corroded concrete structures can be based on complex two- or three-dimensional finite element modelling (Sànchez, Huespe, Oliver, & Toro, 2010). This approach is useful to investigate local stress distribution problems or to analyse the behaviour of structural components and members. However, the involved computational cost and the amount of data to be managed make this approach impractical for the analysis of entire structural systems, such as buildings and bridges. A structural modelling based on beam finite elements represents an effective alternative to obtain sufficiently accurate results at an affordable computational cost.

The formulation of beam finite elements for nonlinear analysis of corroded concrete structures can refer to lumped plasticity models, where nonlinearity is described at the cross-sectional level within critical regions where plastic hinges are expected to occur (Akiyama, Frangopol, & Matsuzaki, 2011; Biondini et al., 2014), or distributed plasticity models, where nonlinearity is defined at the material level in terms of nonlinear constitutive laws of concrete and steel (Biondini et al., 2004a). Compared to lumped plasticity models, beam elements with distributed plasticity usually provide higher accuracy and allow a more detailed modelling of reinforcing steel corrosion and concrete deterioration.

In this paper, the formulation of a three-dimensional deteriorating reinforced concrete (RC) beam finite element for nonlinear analysis of concrete structures under corrosion is presented. The proposed formulation accounts for both mechanical nonlinearity, associated with the constitutive laws of the materials, and geometrical nonlinearity, due to the second-order effects (Biondini, 2004; Biondini, Bontempi, Frangopol, & Malerba, 2004b; Bontempi, Malerba, & Romano, 1995; Malerba, 1998). Damage modelling considers uniform and localised (pitting) corrosion and includes the reduction of cross-sectional area of corroded bars, the reduction of ductility of reinforcing steel and the deterioration of concrete strength due to splitting cracks and spalling of concrete cover (Biondini, 2011; Biondini & Vergani, 2012). In the structural model, these effects are described through damage indices and corrosion can selectively be applied to damaged structural members with a different level of penetration in each reinforcing bar.

The presented developments focus on the deterministic evaluation of the structural effects of prescribed damage patterns and corrosion penetration levels. However, it is worth noting that the proposed formulation can be extended to include the time factor in a lifetime scale by modelling the diffusive process of aggressive agents leading to corrosion initiation and damage propagation, as well as to account for the uncertainty in material and geometrical properties, in the physical models of the deterioration process and in the mechanical and environmental stressors. These extensions allow to incorporate the nonlinear analysis in a probabilistic framework for life-cycle assessment and design of concrete structures exposed to corrosion (Biondini et al., 2004a, 2006).

The beam finite element is validated with reference to the results of experimental tests carried out on RC beams subjected to corrosion of steel reinforcement, including accelerated corrosion (Rodriguez, Ortega, & Casal, 1997) and natural corrosion (Castel, François, & Arliguie, 2000;

Vidal, Castel, & François, 2007). The application to a statically indeterminate RC beam under different corrosion scenarios and increasing levels of corrosion penetration demonstrates the need of a proper evaluation of the global effects of local damage on the overall system performance. Finally, the three-dimensional nonlinear structural analysis of a RC arch bridge under different damage scenarios and corrosion penetration levels is presented. The results highlight the effectiveness of the proposed approach and its application potentialities in the design and assessment of concrete structures exposed to corrosion.

2. Corrosion modelling

Damage modelling includes the reduction of cross-sectional area of corroded bars, the reduction of ductility of reinforcing steel and the deterioration of concrete strength due to splitting cracks, delamination and spalling of the concrete cover. Although corrosion does not affect significantly the yielding strength of reinforcing steel bars (Apostolopoulos & Papadakis, 2008), a moderate reduction of steel strength may be observed for corroded bars with irregular distribution of cross-section loss (Du, Clark, & Chan, 2005). This effect is not considered in the applications studied in this paper, even though it could be easily incorporated in the damage model. The deterioration of steel-concrete bond strength and the effects of the bond-slip of corroded steel bars are also not investigated.

2.1. Reduction of the cross-section of reinforcing bars

The most relevant effect of corrosion is the reduction of the cross-section of the reinforcing steel bars. By denoting p as the corrosion penetration depth, the following dimensionless corrosion penetration index $\delta \in [0;1]$ is introduced (Biondini & Vergani, 2012):

$$\delta = \frac{p}{D_0}, \tag{1}$$

where D_0 is the diameter of the steel bar. The area A_s of a corroded bar can be represented as a function of the corrosion penetration index as follows:

$$A_s(\delta) = [1 - \delta_s(\delta)]A_{s0}, \tag{2}$$

where $A_{s0} = \pi D_0^2/4$ is the area of the undamaged steel bar and $\delta_s = \delta_s(\delta)$ is a dimensionless damage function which provides a measure of cross-section reduction in the range $[0;1]$. The damage function $\delta_s = \delta_s(\delta)$ depends on the corrosion mechanism.

In carbonated concrete without relevant chloride content, corrosion tends to develop uniformly on the

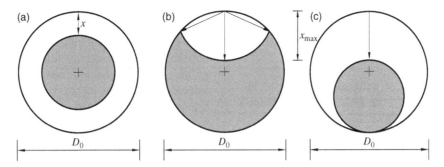

Figure 1. Modelling of cross-section reduction of a corroded reinforcing steel bar: (a) uniform corrosion; (b) pitting corrosion with circular pitting surface; and (c) pitting corrosion with a component of uniform corrosion.

reinforcing bars. Figure 1(a) shows a model of uniform corrosion where the reduction of the bar cross-section depends on the corrosion penetration depth $p = 2x$. This model is described by the following damage function:

$$\delta_s = \delta(2 - \delta). \quad (3)$$

In presence of chlorides, corrosion tends instead to localise (pitting corrosion). This type of corrosion can be characterised by means of a pitting factor R defined as the ratio between the maximum depth x_{max} measured at pit and the average penetration $\bar{x}$ calculated indirectly from the weight loss of the steel bar:

$$R = \frac{x_{max}}{\bar{x}}. \quad (4)$$

Typical values of the pitting factor vary between 4 and 8 for natural corrosion and between 5 and 13 for accelerated corrosion tests (Gonzalez et al., 1995). Regarding spatial distribution, there is no evidence that corrosion develops and propagates at concrete cracks. Instead, it tends to develop randomly where there are imperfections in the passive layer or interface defects (Zhang et al., 2009).

Pits have irregular shape and the cross-section reduction due to pitting corrosion is described through simplified models. As an example, the model shown in Figure 1(b) assumes a circular pitting surface with radius equal to the maximum penetration depth $p = x_{max}$ (Stewart, 2009; Val & Melchers, 1997). The model of pitting corrosion shown in Figure 1(c) is related to a mixed corrosion mechanism with a component of uniform corrosion (Rodriguez, Ortega, Casal, & Diez, 1996). With reference to the maximum penetration depth $p = x_{max}$, this model is described by the same damage function $\delta_s = \delta_s(\delta)$ of uniform corrosion and, with a pitting factor $R = 2$, leads to the same steel mass loss. This type of mixed corrosion mechanism can be found in accelerated corrosion tests, but it is not suitable to represent real cases where uniform or pitting corrosion are predominant (Zhang et al., 2009).

2.2. Reduction of ductility of reinforcing bars

Corrosion may significantly reduce the ductility of reinforcing steel bars. This is mainly related to the spatial variability of the attack penetration. Tensile tests on corroded bars show that for a quite limited mass loss (about 13%), reinforcing steel behaviour may become brittle (Almusallam, 2001). The results of experimental tests reported in Apostolopoulos and Papadakis (2008) indicate that ductility reduction is a function of the cross-section loss. Based on the fitting of available experimental results, the steel ultimate strain ε_{su} of a corroded bar is related to the damage index $\delta_s = \delta_s(\delta)$ as follows (Biondini & Vergani, 2012):

$$\varepsilon_{su} = \begin{cases} \varepsilon_{su0}, & 0 \leq \delta_s < 0.016 \\ 0.1521\delta_s^{-0.4583}\varepsilon_{su0}, & 0.016 < \delta_s \leq 1 \end{cases} \quad (5)$$

where ε_{su0} is the steel ultimate strain of the undamaged bar.

2.3. Effects of corrosion on concrete

The effects of corrosion are not limited to damage of reinforcing steel bars. In fact, particularly in case of uniform corrosion (Val, 2007), the formation of corrosion products may lead to the development of longitudinal cracks in the concrete surrounding the corroded bars and, consequently, to delamination and spalling of the concrete cover. This local deterioration of concrete can effectively be modelled by means of a degradation law of the effective resistant area of the concrete matrix A_c (Biondini et al., 2004a):

$$A_c = [1 - \delta_c(\delta)]A_{c0}, \quad (6)$$

where A_{c0} is the area of undamaged concrete and $\delta_c = \delta_c(\delta)$ is a dimensionless damage function which provides a measure of concrete damage in the range [0;1]. However, in this form, it may be not straightforward to establish a relationship between the damage function δ_c and the corrosion penetration index δ.

Alternatively, the effects of concrete deterioration can be taken into account by modelling the reduction of concrete compression strength f_c due to longitudinal cracking (Biondini & Vergani, 2012):

$$f_c = [1 - \delta_c(\delta)]f_{c0}, \tag{7}$$

where f_{c0} is the strength of undamaged concrete. The reduced concrete strength f_c can be evaluated as follows (Coronelli & Gambarova, 2004):

$$f_c = \frac{f_{c0}}{1 + \kappa(\varepsilon_\perp/\varepsilon_{c0})}, \tag{8}$$

in which κ is a coefficient related to bar diameter and roughness ($\kappa = 0.1$ for medium-diameter ribbed bars), ε_{c0} is the strain at peak stress in compression and $\varepsilon_\perp$ is an average (smeared) value of the tensile strain in cracked concrete at right angles to the direction of the applied stress.

The transversal strain $\varepsilon_\perp$ is evaluated by means of the following relationship:

$$\varepsilon_\perp = \frac{b_f - b_i}{b_i} = \frac{\Delta b}{b_i}, \tag{9}$$

where b_i is the width of the undamaged concrete cross-section and b_f is the width after corrosion cracking. The increase of beam width Δb can be estimated as follows:

$$\Delta b = n_{bars}w, \tag{10}$$

where n_{bars} is the number of steel bars and w is the average crack opening for each bar. Several studies investigated the relationship between the amount of steel corrosion and the crack opening w (Alonso, Andrade, Rodriguez, & Diez, 1998; Zhang et al., 2009). The following empirical model is assumed (Vidal et al., 2004):

$$w = \kappa_w(\delta_s - \delta_{s0})A_{s0}, \tag{11}$$

in which $\kappa_w = 0.0575$ (mm^{-1}) and δ_{s0} is the amount of steel damage necessary for cracking initiation. This damage threshold is evaluated as follows:

$$\delta_{s0} = 1 - \left[1 - \frac{R}{D_0}\left(7.53 + 9.32\frac{c_0}{D_0}\right) \times 10^{-3}\right]^2, \tag{12}$$

where c_0 is the concrete cover.

The crack opening w increases with the expansion of corrosion products up to a critical width w_{cr} which corresponds to the occurrence of delamination and spalling of the concrete cover. Based on experimental evidence, delamination and spalling can occur for crack width in the range $0.1-1.0$ mm (Al-Harthy, Stewart, & Mullard, 2011; Torres-Acosta & Martinez-Madrid, 2003; Vu, Stewart, & Mullard, 2005). In this study, a critical crack width $w_{cr} = 1$ mm is conventionally assumed.

The reduction of concrete strength is generally applied to the entire concrete cover (Coronelli & Gambarova, 2004). However, the longitudinal cracks pattern strongly depends on the arrangement of reinforcing bars. Cracking propagation induced by corrosion should be therefore limited to the zones adjacent to the corroded bars. With reference to a rectangular cross-section, Figure 2 shows a model where the reduction of concrete strength is applied only to a portion of concrete cover surrounding the corroded bars (Biondini & Vergani, 2012).

The cross-section is subdivided in cells and each concrete cell in the neighbourhood of a corroded bar is subjected to damage if at least one of its vertices lies in the intersection of the region surrounding the bar within a radius equal to the cover thickness and the region outside the centroidal circle passing through the bar. This model allows to effectively reproduce the mechanism of spalling of the concrete cover, characterised by inclined fracture planes for wide bar spacing (Figure 2(a)) and parallel fracture planes (i.e. delamination) for closely-spaced bars (Figure 2(b)).

3. Formulation of a deteriorating three-dimensional RC beam finite element

A three-dimensional RC beam finite element, capable to incorporate the effects of corrosion, is presented. The formulation assumes the linearity of the cross-sectional strain field and neglects shear failures and bond-slip of reinforcement. Mechanical nonlinearity, associated with the constitutive laws of the materials, and geometrical nonlinearity, due to the second-order effects, are taken into account (Biondini, 2004; Biondini et al., 2004b; Bontempi et al., 1995; Malerba, 1998).

The local reference system (x', y', z') of the beam finite element and the components of the vector of nodal

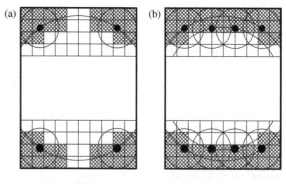

Concrete cell undergoing damage

Figure 2. Definition of the zones of concrete cover undergoing damage: mechanism of cover spalling with (a) inclined fracture planes for wide bar spacing, and (b) parallel fracture planes (delamination) for closely spaced bars.

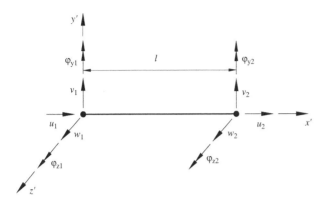

Figure 3. Local reference system and nodal displacements of the beam finite element.

displacements $\mathbf{s}' = [\mathbf{s}_a^{\mathrm{T}} | \mathbf{s}_b^{\mathrm{T}}]^{\mathrm{T}} = [u_1 \ u_2 \ | \ v_1 \ \varphi_{z1} \ v_2 \ \varphi_{z2} \ w_1 \ \varphi_{y1} \ w_2 \ \varphi_{y2}]^{\mathrm{T}}$ are shown in Figure 3. The torsional degrees of freedom are not explicitly included in the formulation because torsion is assumed to be uncoupled from axial and bending deformations. The material $\mathbf{K}_{\mathrm{M}}$ and geometrical $\mathbf{K}_{\mathrm{G}}$ contributions to the element stiffness matrix $\mathbf{K}'$ and the vector of nodal forces $\mathbf{f}'$ equivalent to the applied loads $\mathbf{f}_0 = \mathbf{f}_0(x')$ are obtained by means of the principle of virtual work, in the form of the virtual displacements, and are evaluated by integration over the beam length l as follows:

$$\mathbf{K}' = \mathbf{K}_{\mathrm{M}} + \mathbf{K}_{\mathrm{G}}, \tag{13}$$

$$\mathbf{K}_{\mathrm{M}} = \int_0^l \mathbf{B}^{\mathrm{T}} \mathbf{H} \mathbf{B} \, dx', \tag{14}$$

$$\mathbf{K}_{\mathrm{G}} = \int_0^l N \mathbf{G}^{\mathrm{T}} \mathbf{G} \, dx', \tag{15}$$

$$\mathbf{f}' = \int_0^l \mathbf{N}^{\mathrm{T}} \mathbf{f}_0 \, dx', \tag{16}$$

where x' is the abscissa of the beam axis, $\mathbf{H} = \mathbf{H}(x')$ is the cross-sectional stiffness matrix, $N = N(x')$ is the axial force, $\mathbf{N} = \mathbf{N}(x')$ is the matrix of axial $\mathbf{N}_a = \mathbf{N}_a(x')$ and bending $\mathbf{N}_b = \mathbf{N}_b(x')$ displacement functions, $\mathbf{B} = \mathbf{B}(x')$ and $\mathbf{G} = \mathbf{G}(x')$ are the corresponding compatibility matrices:

$$\mathbf{N} = \begin{bmatrix} \mathbf{N}_a & \mathbf{0} \\ \mathbf{0} & \mathbf{N}_b \end{bmatrix}, \tag{17}$$

$$\mathbf{B} = \begin{bmatrix} \dfrac{\partial \mathbf{N}_a}{\partial x'} & \mathbf{0} \\ \mathbf{0} & \dfrac{\partial^2 \mathbf{N}_b}{\partial x'^2} \end{bmatrix}, \tag{18}$$

$$\mathbf{G} = \begin{bmatrix} \mathbf{0} & \dfrac{\partial \mathbf{N}_b}{\partial x'} \end{bmatrix}. \tag{19}$$

The matrix $\mathbf{N} = \mathbf{N}(x')$ is defined by adopting the Hermitian shape functions of a linear elastic beam having uniform cross-sectional stiffness and loaded only at its ends. However, due to material nonlinearity, the cross-sectional stiffness distribution along the beam is non-uniform even for prismatic members with uniform reinforcement. The stiffness matrix $\mathbf{H} = \mathbf{H}(x')$ is computed by integration over the area of the composite cross-section or by assembling the contributions of both concrete $\mathbf{H}_c = \mathbf{H}_c(x')$ and reinforcing steel $\mathbf{H}_s = \mathbf{H}_s(x')$:

$$\mathbf{H} = \mathbf{H}_c + \mathbf{H}_s, \tag{20}$$

$$\mathbf{H}_c = \int_{A_c} \bar{E}_c \mathbf{b}^{\mathrm{T}} \mathbf{b} dA, \tag{21}$$

$$\mathbf{H}_s = \sum_m \bar{E}_{\mathrm{sm}} \mathbf{b}_m^{\mathrm{T}} \mathbf{b}_m A_{\mathrm{sm}}, \tag{22}$$

where the symbol 'm' refers to the mth reinforcing bar located at point (y_m', z_m') in the centroidal principal reference system (y', z') of the cross-section, $\bar{E}_c = \bar{E}_c(x', y', z')$ and $\bar{E}_{\mathrm{sm}} = \bar{E}_{\mathrm{sm}}(x')$ are the secant moduli of the materials, and $\mathbf{b}(y', z') = [1 \ -y' \ z']$ is the linear compatibility operator of the cross-sectional strain field. In this study, the stress-strain law of concrete is described by the Saenz's law in compression and a bilinear softening model in tension. For steel, a bilinear elastic-plastic model in both tension and compression is assumed.

The damage effects associated with the corrosion penetration index δ are included in this formulation by assuming $A_s = A_s(\delta)$ and $\varepsilon_{\mathrm{su}} = \varepsilon_{\mathrm{su}}(\delta)$ for the steel bars and $f_c = f_c(\delta)$ for the concrete to obtain the deteriorating stiffness matrices $\mathbf{H} = \mathbf{H}(x', \delta)$ and $\mathbf{K}' = \mathbf{K}'(\delta)$ (Biondini, 2011; Biondini & Vergani, 2012). It is worth noting that corrosion can selectively be applied to damaged structural elements with a different level of corrosion penetration for each reinforcing bar.

The quantities which define the characteristics of the beam finite element are evaluated by numerical integration. The volume of the element is subdivided in prismatic isoparametric sub-domains having quadrilateral cross-section, and each sub-domain is replaced by a grid of points whose location depends on the adopted integration rule (Bontempi et al., 1995). In particular, depending on the function to be integrated and the geometry of the problem, the Gauss-Legendre and the Gauss-Lobatto schemes are used.

Finally, by assembling the stiffness matrix $\mathbf{K}'$ and the vectors of nodal forces $\mathbf{f}'$ for all members in a global reference system (x, y, z), the equilibrium of the structure can be expressed as follows:

$$\mathbf{Ks} = \mathbf{f}, \qquad (23)$$

where $\mathbf{K} = \mathbf{K}(\mathbf{s}, \delta)$ is the overall stiffness matrix, $\mathbf{s} = \mathbf{s}(\delta)$ is the vector of nodal displacements and $\mathbf{f}$ is the vector of applied nodal forces. This nonlinear equation system is solved numerically by means of the secant method for prescribed values of the corrosion penetration index δ.

4. Validation of the RC beam finite element

The RC beam finite element is validated with reference to experimental tests carried out on beams with corroded reinforcement.

4.1. Beams subjected to accelerated corrosion tests

The experimental tests reported by Rodriguez et al. (1997) are considered. The validation procedure refers to the two type of beams shown in Figure 4. The beams were cast adding calcium chloride to the mixing water, subjected to an accelerated corrosion process with a current density of $100\,\mu\text{A/cm}^2$ and finally loaded up to failure.

The beams type 11, with a lower reinforcement ratio, showed a flexural failure with rupture of the tensile bars. For beams type 31, with a higher reinforcement ratio, a crushing failure of concrete in compression occurred. For these beams, the average corrosion penetration depth $\bar{x}$, as calculated indirectly from weight loss of the steel bars, and the maximum penetration depth $x_{\max}$ measured at pits are listed in Table 1. The mechanical properties of concrete and steel are listed in Tables 2(a) and 3(a), respectively.

The structural symmetry of the model is exploited and the half beam is discretised into six finite elements with five Gauss-Lobatto sampling points in each element. The cross-

Table 1. Beams under accelerated corrosion: average corrosion penetration depth, as calculated indirectly from weight loss of the corroded steel bars, and maximum penetration depth measured at pits (value in parenthesis). Adapted from Rodriguez et al. (1997).

	Corrosion penetration depth (mm)	
Beam type	Tensile bars	Compressive bars
114	0.45 (1.1)	0.52
115	0.36 (1.0)	0.26
116	0.71 (2.1)	0.48
313	0.30 (1.3)	0.20
314	0.48 (1.5)	0.26
316	0.42 (1.8)	0.37

section is subdivided into 300 isoparametric sub-domains, with a grid of 3×3 Gauss-Lobatto sampling points in each domain. The cross-sectional area of the corroded bars is computed by assuming the model of pitting corrosion shown in Figure 1(c) with maximum penetration depth $x_{\max}$ for the bars in tension, and uniform corrosion with penetration depth $x = \bar{x}$ for the bars in compression (Table 1).

A comparison between the results of the nonlinear analyses and the results of the tests is shown in Figure 5 in terms of load versus midspan displacement diagrams. The good agreement between experimental and numerical results demonstrates the accuracy of the proposed formulation, as well as its capability to capture the main corrosion degradation mechanisms. With this regard, Figure 6 shows the cross-sectional contour maps of concrete damage of the beams 116 and 314. These maps highlight the important role of the reinforcement layout for the propagation of damage in the concrete cover. In fact, for the beams type 11, the corrosion of the steel bars located at the cross-section corners leads to a localised spalling along inclined fracture planes (Figure 6(a)). For the beams type 31,

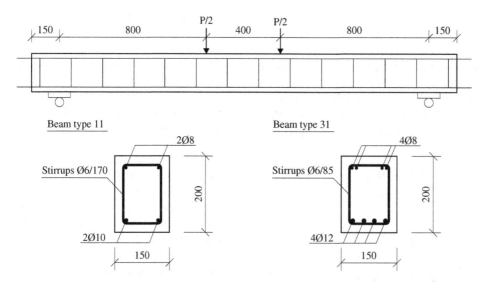

Figure 4. Experimental tests on beams subjected to accelerated corrosion: geometrical dimensions (mm), testing scheme and characteristics of the cross-sections for beams type 11 and type 31. Adapted from Rodriguez et al. (1997).

Table 2. Mechanical properties of concrete: (a) beams under accelerated corrosion, (b) beams under natural corrosion, (c) statically indeterminate beam, (d) arch bridge.

	Case study	Element type	f_c (MPa)	f_{ct} (MPa)	E_{c0} (GPa)
(a)	Beams under accelerated corrosion tests	Beam 111	50.0	4.1	37.3
		Beams 114-115-116	34.0	3.1	33.8
		Beam 311	49.0	4.1	37.1
		Beams 313-314-316	37.0	3.2	34.5
(b)	Beams under natural corrosion tests	Beam B1T	65.3	0.0	36.3
		Beam B1CL	63.4	0.0	35.0
(c)	Statically indeterminate beam	Beam	35.0	3.2	34.0
(d)	Arch bridge	Arches & Deck	35.0	3.2	34.0

Table 3. Mechanical properties of steel: (a) beams under accelerated corrosion, (b) beams under natural corrosion, (c) statically indeterminate beam, (d) arch bridge.

	Case study	Element type	f_{sy} (MPa)	f_{su} (MPa)	E_s (GPa)	ε_{su}
(a)	Beams under accelerated corrosion tests	Bars ø8	615	673	210	0.03
		Bars ø10	575	655	210	0.03
		Bars ø12	585	673	210	0.03
(b)	Beams under natural corrosion tests	Bars (B1T & B1CL)	500	500	210	0.06
(c)	Statically indeterminate beam	Bars	440	440	210	0.01
(d)	Arch bridge	Bars	440	440	210	0.01
		IPN I-beams	275	275	210	0.10
		Ties	1600	1600	200	0.10

a similar pattern of localised damage would occur at the top side of the cross-section for the steel bars arranged in pairs at the corners. The corrosion of the four closely-spaced steel bars at the bottom side leads instead to the delamination and spalling of the entire concrete cover (Figure 6(b)).

However, it is worth noting that the lower values of penetration depth of the bars in compression may be not sufficient to induce the spalling of concrete cover. This is illustrated in Figure 7, which shows the comparison between experimental and numerical results obtained with and without the activation of the spalling of the concrete cover at the top side of the cross-section for the beam 314. If cover spalling is activated, the numerical model underestimates the stiffness associated with the initial branch of the load–displacement diagram. As shown in Figure 7, these results are also in good agreement with the numerical results obtained by Sànchez et al. (2010). On the contrary, if cover spalling is not activated, the numerical results of the finite element analysis reproduce the initial branch of the load-displacement diagram with good accuracy.

4.2. Beams subjected to natural corrosion tests

The experimental tests reported by Castel et al. (2000) and Vidal et al. (2007), concerning beams subjected to natural corrosion, are considered. The tested beams were exposed to a salt fog (35 g/l of NaCl, corresponding to the salt concentration of sea water) and subjected to a three-point flexure up to collapse on a clear span of 2.8 m. The beams have rectangular cross-section 150 mm × 280 mm. They are reinforced with two bars with diameter $\phi = 12$ mm at the bottom side in tension and two bars with diameter $\phi = 6$ mm at the top side in compression, with a concrete cover $c_0 = 10$ mm. Based on the available information, the mechanical properties listed in Tables 2(b) and 3(b) for concrete and steel, respectively, are assumed. From the set of tested specimens, a non corroded control beam (B1T) and a beam after 14 years of exposure (B1CL) are considered for the validation procedure.

In the structural model the beam is discretised into six finite elements with five Gauss-Lobatto sampling points in each element. The cross-section is subdivided into 200 isoparametric sub-domains, with a grid of 3 × 3 Gauss-Lobatto sampling points in each domain. The nonlinear analysis of the corroded beam is performed by assuming a steel cross-section loss of 22%, which corresponds to the average value of the maximum cross-section losses measured for the two tensile reinforcing steel bars in the critical region at midspan (Castel et al., 2000). Figure 8 shows the comparison between experimental and numerical results in terms of load versus midspan displacement. The comparison confirms that the proposed formulation is able to reproduce with good accuracy the structural response of noncorroded and corroded structures and to provide a very accurate estimation of both the load carrying capacity and the maximum displacement at failure.

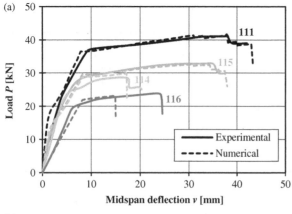

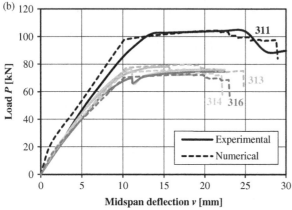

Figure 5. Beams under accelerated corrosion: comparison between experimental and numerical results for (a) beams type 11 and (b) beams type 31.

5. Applications

5.1. *Statically indeterminate beam*

The RC beam shown in Figure 9 is considered. The material properties of concrete and steel are listed in Tables 2(c) and 3(c), respectively. The structural response of the beam is investigated by assuming uniform corrosion of the longitudinal steel reinforcement localised over a length of 50 cm from the clamped end. Figure 10(a),(b) shows the deformed shape and the bending moment diagram,

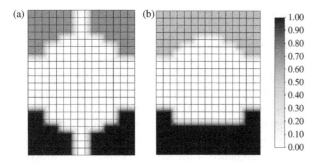

Figure 6. Beams under accelerated corrosion: cross-sectional contour maps of concrete damage of (a) beam 116 and (b) beam 314.

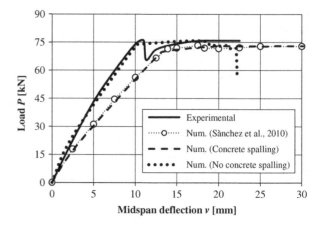

Figure 7. Beams under accelerated corrosion: comparison between experimental and numerical results, obtained with and without the activation of the spalling of the concrete cover, for the beam 314.

respectively, for increasing values of the corrosion penetration index δ. These results highlight the effects of localised damage on the global structural response. In the case studied, localised damage involves both an increase of the overall deformability of the beam and, due to statical indeterminacy, a redistribution of the internal stresses. In particular, the structural response of the damaged beam tends to reproduce the behaviour of a simply supported beam (dashed curve in Figure 10(a),(b)) when the corrosion penetration index approaches the limit value $\delta = 1$. This indicates that for statically indeterminate structures, the internal stresses under constant loading may change over time as a consequence of the time evolution of damage.

The role of the damage exposure scenario is also investigated by considering three cases of uniform corrosion distributed over the beam length, with corrosion of (I) top reinforcement only, (II) bottom reinforcement only and (III) both top and bottom reinforcement. Figure 11(a),(b) shows, for each case studied, the maximum beam deflection and the bending moment at the clamped beam end, respectively, versus the corrosion penetration index. It is worth noting that a sudden change of slope of these curves is related to damage thresholds leading to the first yielding of steel reinforcement. Figure 11(a) shows that corrosion leads to an increase of the beam deformability. As expected, the maximum increase of beam deflection is achieved for case (III), with corrosion of both top and bottom reinforcement. Moreover, the comparison of case (I) and case (II) highlights that corrosion of bottom reinforcement leads to higher values of beam deflection with respect to corrosion of top reinforcement. In terms of stress redistribution, the bending moment at the clamped beam end decreases for case (I), with corrosion of top reinforcement, and increases for case (II), with corrosion of bottom reinforcement, as the corrosion penetration index increases. For case (III), with corrosion of top and bottom reinforcement, the bending moment at the clamped beam

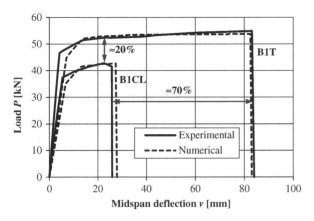

Figure 8. Beams under natural corrosion: comparison between experimental and numerical results.

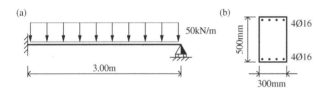

Figure 9. RC beam exposed to corrosion: (a) structural scheme and loading and (b) geometry of the cross-section and reinforcement layout.

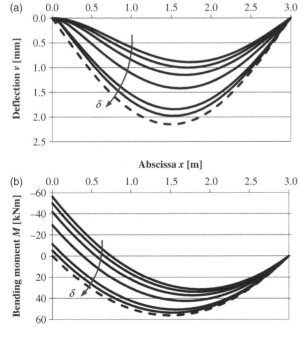

Figure 10. Statically indeterminate beam: (a) deformed shape and (b) bending moment diagram for different values of corrosion penetration depth ($\delta = 0$, 0.2, 0.4, 0.6, 0.8, 1.0), compared with the limit case of undamaged simply supported beam (dashed curve).

end does not change for corrosion levels up to the threshold value $\delta = 0.42$ associated with the first yielding of top reinforcement and decreases for increasing levels of corrosion $\delta > 0.42$ up to collapse.

Based on the results of this application, it is clear that design for durability of concrete structures cannot be limited to simplified prescriptions associated with the quality of the materials and structural details such as the minimum concrete cover, but also needs to rely on nonlinear structural analysis frameworks capable to take into account the global effects of local damage phenomena on the overall system performance.

5.2. Arch bridge

The structural response of a Nielsen type RC arch bridge with the deck suspended by inclined steel ties is investigated under different damage scenarios and corrosion penetration levels. Many structures of the same type were built in Europe in the Thirties, with spans varying between 50 and 140 m. Figure 12 shows the structural scheme and the main dimensions of the arch bridge, as well as the

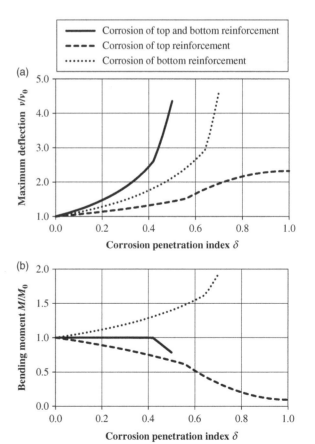

Figure 11. Statically indeterminate beam: damaged/undamaged ratio of (a) maximum deflection and (b) bending moment at the clamped end of the corroded beam versus the corrosion penetration index.

geometry, dimensions and reinforcement layout of the structural members. The three-dimensional structural model of the bridge is shown in Figure 13.

The span of the arch bridge is 90 m, subdivided into 12 sub-spans by the anchoring points of the suspension ties. The rise of the arches is 13 m. The static scheme allows the elimination of the horizontal thrust that is carried by means of two IPN 600 European standard steel I-beams inserted into the concrete main beams of the deck. The bridge deck is stiffened by transversal ribs with rectangular cross-section 250 mm × 640 mm and 3.75 m

spacing. The arches are connected by seven transversal T-beams with top flange 700 mm × 200 mm and beam web 200 mm × 400 mm. The beams are located in correspondence with the anchoring points of some ties and at midspan, as shown in Figure 13. Further details about the geometry and reinforcement layout of the transversal beams of both deck and arches can be found in Vergani (2010). The suspension ties are steel bars with diameter $\phi = 70$ mm. The material properties of concrete and different types of steel are listed in Tables 2(d) and 3(d), respectively.

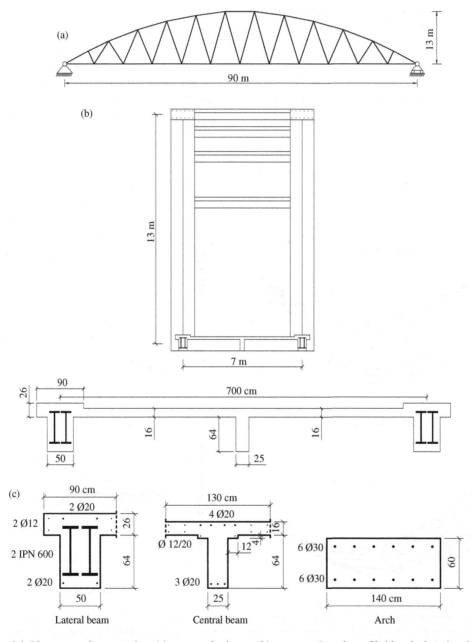

Figure 12. RC arch bridge exposed to corrosion: (a) structural scheme, (b) cross-section view of bridge deck and arches, and (c) cross-sections of the deck's main beams and of the arches.

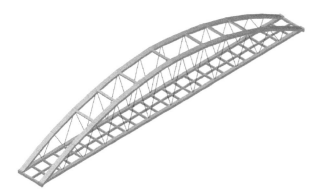

Figure 13. View of the three-dimensional structural model of the arch bridge.

The bridge deck is modelled by means of a grillage formed by the three longitudinal main beams and 25 transversal beams. The structural response of the bridge is analysed under the self weight $g_1 = 8\,\text{kN/m}^2$ and $g_2 = 21\,\text{kN/m}$ of the deck and the arches, respectively, and a live load q uniformly distributed over the deck area and increased up to collapse. The load associated with each grid unit $3.50\,\text{m} \times 3.75\,\text{m}$ of the deck grillage is applied to the corresponding edge beam elements.

The effects of corrosion on the bridge structural performance are investigated at the system level by considering three damage scenarios with corrosion applied separately to deck, arches and ties. The results of this type of analyses allow to clarify which are the structural elements that more significantly affect the structural performance if exposed to corrosion. This information could provide the basis to establish a hierarchy of protection techniques, as well as to plan proper maintenance actions and repair interventions.

The structural analyses are carried out by assuming uniform corrosion for prescribed values of corrosion

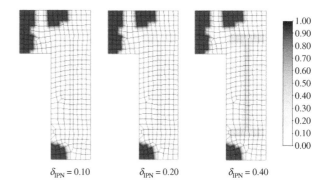

Figure 15. Arch bridge: contour maps of concrete damage of the lateral beams of the deck (half cross-section).

penetration depth for reinforcing steel bars, IPN I-beams and steel ties. For steel bars and steel ties, the damage level is described by using the corrosion penetration index δ. For the IPN I-beams, the level of corrosion is more conveniently described in terms of cross-section loss as follows:

$$\delta_{\text{IPN}} = 1 - \frac{A_{\text{IPN}}}{A_{\text{IPN},0}}, \qquad (24)$$

where A_{IPN} is the area of the corroded steel beam, and $A_{\text{IPN},0}$ is the corresponding value in the undamaged state.

Figure 14 depicts the load versus the maximum deflection of the lateral main beams for different damage levels of the bridge deck ($\delta = \delta_{\text{IPN}} = 0, 0.05, 0.10, 0.20, 0.40$). The results indicate that this damage scenario involves an increase of deformability and a significant decrease of load carrying capacity, with small changes in terms of displacement ductility. It is interesting to note that even for a limited damage of the IPN I-beams, the corrosion of the reinforcing steel bars may lead to the

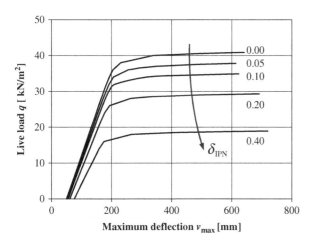

Figure 14. Arch bridge: load versus maximum deflection of the lateral main beams for different damage levels of the bridge deck ($\delta = \delta_{\text{IPN}} = 0, 0.05, 0.10, 0.20, 0.40$).

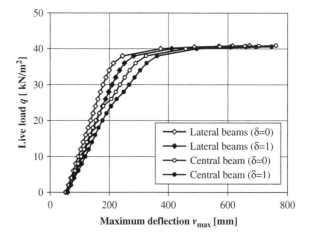

Figure 16. Arch bridge: load versus maximum deflection of the lateral and central main beams of the deck for the cases of undamaged bridge ($\delta = 0$) and bridge with damaged arches with complete corrosion penetration ($\delta = 1$).

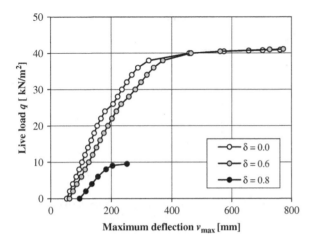

Figure 17. Arch bridge: load versus maximum deflection of the central main beam of the deck for the cases of undamaged bridge ($\delta = 0$) and bridge with damaged suspension steel ties with corrosion penetration depth $\delta = 0.6$ and $\delta = 0.8$.

spalling of the concrete cover, as shown by the cross-sectional damage contour maps in Figure 15.

The effects of corrosion applied to the arches are presented in Figure 16 in terms of load versus maximum deflection of the lateral and central beams of the deck for the cases of undamaged bridge ($\delta = 0$) and bridge with damaged arches with complete corrosion penetration ($\delta = 1$). The results show that the deterioration of the arches, which also involves the spalling of the concrete cover, does not significantly affect the structural performance.

Similar results are obtained when corrosion damage is applied to the suspension steel ties with a corrosion penetration depth up to $\delta = 0.60$, as shown in Figure 17. However, further cross-section reductions of the steel ties involve a remarkable deterioration of both load carrying

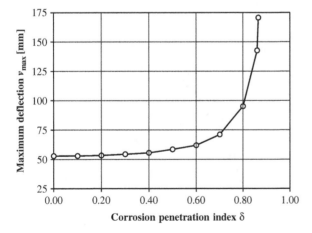

Figure 18. Arch bridge: maximum deflection of the central main beam of the deck under dead load ($q = 0$) versus the corrosion penetration index for the case of bridge with damaged suspension steel ties.

capacity and ductility of the bridge structure, as shown in Figure 17 for $\delta = 0.80$, and lead to structural collapse under dead load for a corrosion penetration depth $\delta = 0.87$. Figure 18 shows the effect of corrosion of the suspension steel ties on the maximum deflection of the central main beam of the deck under dead load ($q = 0$) for increasing values of corrosion penetration depth up to collapse.

6. Conclusions

A three-dimensional RC beam finite element for nonlinear analysis of concrete structures exposed to corrosion has been presented. The proposed formulation allows to model the damage effects of uniform and pitting corrosion in terms of reduction of cross-sectional area of corroded bars, reduction of ductility of reinforcing steel, deterioration of concrete strength and spalling of concrete cover. The beam finite element has been validated with reference to experimental tests carried out on RC beams subjected to accelerated and natural corrosion. The comparison of experimental and numerical results demonstrated the accuracy of the proposed approach and its capability to reproduce the effects of local corrosion damage on the overall structural response. Moreover, the results of the validation process highlighted the importance of a proper modelling of the delamination and spalling of the concrete cover in compression to accurately predict the initial branch of the load–displacement relationships.

The application to a statically indeterminate RC beam under corrosion showed that design for durability of concrete structures cannot be limited to simplified prescriptions on the quality of the materials and structural detailing, but also needs structural analysis methods capable to take into account the global effects of local damage phenomena on the overall performance of the structure. Finally, the three-dimensional nonlinear structural analysis of a RC arch bridge under different damage scenarios and corrosion penetration levels highlighted the effectiveness and application potentialities of the proposed formulation. In particular, it has been shown how damage can selectively be applied to identify the structural components and members that more significantly affect the system performance when they are exposed to corrosion. This information could provide the basis to establish a hierarchy of protection techniques, as well as to plan proper maintenance actions and repair interventions.

This study focused on the deterministic evaluation of the effects associated with prescribed exposure scenarios and corrosion penetration levels. However, the proposed approach can be extended to account for the uncertainty involved in the problem in probabilistic terms and to include the time factor in a lifetime scale by modelling the diffusive process of aggressive agents leading to corrosion initiation and damage propagation, as described in Biondini et al. (2004a, 2006).

Further developments are needed to include the corrosion of the stirrups and the bond strength deterioration in the damage model and to incorporate the effects of shear behaviour, bond-slip of reinforcement and cyclic loading in the formulation of the deteriorating beam finite element. These factors can be particularly relevant for seismic design and assessment of corroded concrete structures (Akiyama et al., 2011; Biondini et al., 2014).

References

Akiyama, M., Frangopol, D.M., & Matsuzaki, H. (2011). Life-cycle reliability of RC bridge piers under seismic and airborne chloride hazards. *Earthquake Engineering and Structural Dynamics, 40*, 1671–1687.

Al-Harthy, A.S., Stewart, M.G., & Mullard, J. (2011). Concrete cover cracking caused by steel reinforcement corrosion. *Magazine of Concrete Research, 63*, 655–667.

Almusallam, A.A. (2001). Effect of degree of corrosion on the properties of reinforcing steel bars. *Construction and Building Materials, 15*, 361–368.

Alonso, C., Andrade, J., Rodriguez, J., & Diez, J.M. (1998). Factors controlling cracking of concrete affected by reinforcement corrosion. *Materials and Structures, 31*, 435–441.

ASCE (2013). *Report card for America's infrastructure.* Reston, VA: American Society of Civil Engineers.

Apostolopoulos, C.A., & Papadakis, V.G. (2008). Consequences of steel corrosion on the ductility properties of reinforcement bar. *Construction and Building Materials, 22*, 2316–2324.

Bhargava, K., Gosh, A.K., Mori, Y., & Ramanujam, S. (2007). Corrosion-induced bond strength degradation in reinforced concrete – Analytical and empirical models. *Nuclear Engineering and Design, 237*, 1140–1157.

Biondini, F. (2004). A three-dimensional finite beam element for multiscale damage measure and seismic analysis of concrete structures. *13th World Conference on Earthquake Engineering (13 WCEE)*, Vancouver, BC, Canada, 1–6 August, Paper no. 2963.

Biondini, F. (2011). Cellular automata simulation of damage processes in concrete structures. In Y. Tsompanakis & B.H. V. Topping (Eds.), *Soft computing methods for civil and structural engineering*, Chapter 10 (pp. 229–264). Stirlingshire: Saxe-Coburg.

Biondini, F., Bontempi, F., Frangopol, D.M., & Malerba, P.G. (2004a). Cellular automata approach to durability analysis of concrete structures in aggressive environments. *ASCE Journal of Structural Engineering, 130*, 1724–1737.

Biondini, F., Bontempi, F., Frangopol, D.M., & Malerba, P.G. (2004b). Reliability of material and geometrically nonlinear reinforced and prestressed concrete structures. *Computers & Structures, 82*, 1021–1031.

Biondini, F., Bontempi, F., Frangopol, D.M., & Malerba, P.G. (2006). Probabilistic service life assessment and maintenance planning of concrete structures. *ASCE Journal of Structural Engineering, 132*, 810–825.

Biondini, F., Camnasio, E., & Palermo, A. (2014). Lifetime seismic performance of concrete bridges exposed to corrosion. *Structure and Infrastructure Engineering, 10*, 880–900.

Biondini, F., & Frangopol, D.M. (2008). Probabilistic limit analysis and lifetime prediction of concrete structures. *Structure and Infrastructure Engineering, 4*, 399–412.

Biondini, F., & Vergani, M. (2012). Damage modeling and nonlinear analysis of concrete bridges under corrosion. *Sixth International Conference on Bridge Maintenance, Safety and Management* (IABMAS 2012), Stresa, Italy, 8–12 July. In F. Biondini & D.M. Frangopol (Eds.), *Bridge Maintenance, Safety, Management, Resilience and Sustainability.* London: CRC Press/Balkema, Taylor & Francis Group.

Bontempi, F., Malerba, P.G., & Romano, L. (1995). Formulazione diretta secante dell'analisi non lineare di telai in C.A./ C.A.P. [Secant formulation of nonlinear analysis of RC/PC frames]. In *Studi e Ricerche*, Graduate School for Concrete Structures 'F.lli Pesenti', Politecnico di Milano, Italy, 16, (in Italian) (pp. 351–386).

Castel, A., François, R., & Arliguie, G. (2000). Mechanical behaviour of corroded reinforced concrete beams – Part 1: Experimental study of corroded beams. *Materials and Structures, 33*, 539–544.

CEB (1992). *Durable concrete structures – Design guide.* Comité Euro-international du Béton, Bulletin d'information no. 183. London: Thomas Telford.

Coronelli, D., & Gambarova, P. (2004). Structural assessment of corroded reinforced concrete beams: Modeling guidelines. *ASCE Journal of Structural Engineering, 130*, 1214–1224.

Du, Y.G., Clark, L.A., & Chan, A.H.C. (2005). Residual capacity of corroded reinforcing bars. *Magazine of Concrete Research, 57*, 135–147.

El Maaddawy, T.A., & Soudki, K. (2007). A model for prediction of time from corrosion initiation to corrosion cracking. *Cement and Concrete Composites, 29*, 168–175.

Ellingwood, B.R. (2005). Risk-informed condition assessment of civil infrastructure: State of practice and research issues. *Structure and Infrastructure Engineering, 1*, 7–18.

fib (2006). *Model code for service life design.* Bulletin no. 34. Lausanne: fédération internationale du béton/The International Federation for Structural Concrete.

Frangopol, D.M. (2011). Life-cycle performance, management, and optimization of structural systems under uncertainty: Accomplishments and challenges. *Structure and Infrastructure Engineering, 7*, 389–413.

Frangopol, D.M., & Ellingwood, B.R. (2010). Life-cycle performance, safety, reliability and risk of structural systems, Editorial. *Structure Magazine*, Joint Publication of NCSEA. CASE, SEI. Reedsburg, WI: C_3Ink.

Gonzalez, J.A., Andrade, C., Alonso, C., & Feliu, S. (1995). Comparison of rates of general corrosion and maximum pitting penetration on concrete embedded steel reinforcement. *Cement and Concrete Research, 25*, 257–264.

Li, C.Q., Zheng, J.J., Lawanwisut, W., & Melchers, R.E. (2007). Concrete delamination caused by steel reinforcement corrosion. *ASCE Journal of Materials in Civil Engineering, 19*, 591–600.

Lundgren, K. (2007). Effect of corrosion on the bond between steel and concrete: An overview. *Magazine of Concrete Research, 59*, 447–461.

Malerba, P.G. (Ed.). (1998). *Analisi limite e non lineare di strutture in calcestruzzo armato* [Limit and nonlinear analysis of reinforced concrete structures]. Udine (in Italian): International Centre for Mechanical Sciences (CISM).

NCHRP (2006). *Manual on service life of corrosion-damaged reinforced concrete bridge superstructure elements.* National Cooperative Highway Research Program, Report 558. Washington, DC: Transportation Research Board.

Pantazopoulou, S.J., & Papoulia, K.D. (2001). Modeling cover-cracking due to reinforcement corrosion in RC structures. *ASCE Journal of Engineering Mechanics, 127*, 342–351.

Rodriguez, J., Ortega, L.M., & Casal, J. (1997). Load carrying capacity of concrete structures with corroded reinforcement. *Construction and Building Materials, 11,* 239–248.

Rodriguez, J., Ortega, L.M., Casal, J., & Diez, J.M. (1996). Corrosion of reinforcement and service life of concrete structures. In C. Sjostrom (Ed.), *Durability of building materials and components* (Vol. 1, pp. 117–126). London: E & FN Spon.

Sànchez, P.J., Huespe, A.E., Oliver, J., & Toro, S. (2010). Mesoscopic model to simulate the mechanical behavior of reinforced concrete members affected by corrosion. *International Journal of Solids and Structures, 47,* 559–570.

Stewart, M.G. (2009). Mechanical behaviour of pitting corrosion of flexural and shear reinforcement and its effect on structural reliability of corroding RC beams. *Structural Safety, 31,* 19–30.

Torres-Acosta, A.A., & Martinez-Madrid, M. (2003). Residual life of corroding reinforced concrete structures in marine environment. *ASCE Journal of Materials in Civil Engineering, 15,* 344–353.

Val, D.V. (2007). Deterioration of strength of RC beams due to corrosion and its influence on beam reliability. *ASCE Journal of Structural Engineering, 133,* 1297–1306.

Val, D.V., & Melchers, R.E. (1997). Reliability of deteriorating RC slab bridges. *ASCE Journal of Structural Engineering, 123,* 1638–1644.

Vergani, M. (2010). *Modellazione del degrado di strutture in calcestruzzo armato soggette a corrosione* [Damage modeling of reinforced concrete structures subjected to corrosion] (Graduate thesis), Milan (in Italian): Politecnico di Milano.

Vidal, T., Castel, A., & Francois, R. (2004). Analyzing crack width to predict corrosion in reinforced concrete. *Cement and Concrete Research, 34,* 165–174.

Vidal, T., Castel, A., & François, R. (2007). Corrosion process and structural performance of a 17 years old reinforced concrete beam stored in chloride environment. *Cement and Concrete Research, 37,* 1551–1561.

Vu, K., Stewart, M.G., & Mullard, J. (2005). Corrosion-induced cracking: experimental data and predictive models. *ACI Structural Journal, 102,* 719–726.

Zhang, R., Castel, A., & François, R. (2009). Concrete cover cracking with reinforcement corrosion of RC beams during chloride-induced corrosion process. *Cement and Concrete Research, 39,* 1077–1086.

Wind tunnel: a fundamental tool for long-span bridge design

Giorgio Diana, Daniele Rocchi and Marco Belloli

An overview of wind tunnel activities and methodologies to support the design of long-span suspension bridges is proposed. The most important aspects of the wind-bridge interaction are investigated considering the aerodynamic phenomena affecting the different parts of the bridge (mainly deck and towers). The experimental activities and results are proposed in the framework of a synergic approach between numerical and experimental methodologies that represent the common practice in defining the full scale aeroelastic behaviour of the bridge starting from scaled reproduction of the wind-bridge interaction. Static and dynamic wind loads, aeroelastic stability, vortex-induced vibrations will be investigated.

Introduction

The definition 'long span bridges' is usually related to bridges with the main span length of the order of 1000–1500 m. For these bridges, the flexibility is high and the first natural frequencies are of the order of 0.1 Hz or lower. Bridges with span lengths over 1.5 km are classified as 'very long span bridges' and their natural frequencies decrease in inverse proportion of the span length. For the 3.3 km span of the Messina bridge, the first natural frequency is 0.03 Hz. These bridges are really sensitive to the wind action, and for them, wind becomes the major problem affecting the overall design.

The maximum span length of a cable-stayed bridge is 1.1 km: all bridges with longer spans are suspension bridges. Currently, the longest bridge is the Akashi in Japan, with a main span of 1990 m. Considering the bridges at the stage of the detailed final design, the longest one is the Messina strait bridge in Italy. For these very long span bridges, as already said, wind plays the most important role in bridge design. In the following, we will make reference mainly to suspension bridges.

Many problems must be faced to guarantee bridge performance to wind action; the most crucial are:

(1) Static load due to the average component of the wind blowing on the structure.
(2) Dynamic load due to the incoming wind turbulence.
(3) Instability of the bridge.
(4) Vortex shedding on the tower, deck and cables.

The analysis of these problems required the development of special models. These models need tests in wind tunnel to identify the aerodynamic and aeroelastic parameters of the different bridge components. Various types of wind tunnel tests are needed:

- for sectional/aeroelastic models of the deck, tower and cables;
- for aeroelastic models of the full bridge: this is the final check.

The analytical methods and the experimental tests in the wind tunnel are strongly correlated and some of the wind tunnel tests are dedicated to the identification of the aerodynamic and aeroelastic parameters used in the computation programmes.

This paper is mainly aimed at the description of the types of tests generally preformed in a wind tunnel at the design stage, in order to analyse the bridge response to wind action and guarantee the correct performances. For a better understanding of the overall problem, the analytical methods correlated to the wind tunnel tests are also briefly described. The paper is organised in the following sections: (i) introduction, (ii) description of the main problems related to the wind action, (iii) analytical methods for the analysis of the bridge response to the wind, (iv) wind tunnel tests: deck, tower as well as the full bridge and (v) conclusions.

Description of the main problems related to the wind action

Static load

Static loads due to wind aerodynamic forces are very important for a long-span bridge. They are due to the average

Figure 1. Comparison of Akashi (left) and Messina (right) horizontal deflection due to wind in the wind tunnel. Equivalent wind speed = 60 m/s. The Akashi photo is courtesy of Honshu-Shikoku Bridge Expressway Company Limited – Japan.

wind component and they condition all bridge design processes. These loads are applied to towers, deck and main cables. The drag load of the deck plays the most important role. The static wind load on the deck is transferred through the hangers to the main cables because the deck lateral stiffness is negligible in comparison with the main cables, in the case of very long span bridges. The deck lateral stiffness is decreasing as a function of the length with a power of 3 laws, while cable stiffness is decreasing linearly with the length. Therefore, the overall static wind load (deck + - cables) is applied to the top of the towers, producing a very large flexural moment on the tower. This is one of the most important loads affecting the tower design.

From this point of view, it is clear that the deck drag must be kept as small as possible, in order to reduce the load at the top of the towers. To better appreciate the problem, we can compare the maximum static horizontal deflection of the Akashi bridge with that of the Messina bridge (see Figure 1). Figure 1 illustrates the deflections of the Akashi and Messina bridge aeroelastic models during wind tunnel tests at an equivalent full scale wind speed of 60 m/s. The maximum deflection for Messina at mid span is around 10 m, compared with approximately 30 m for Akashi. These results are mainly due to the larger deck drag of Akashi with respect to that of Messina (see Figure 11). We can therefore summarise this part of the analysis by stating that the deck drag coefficient is one of the parameters that has to be minimised for very long suspension bridges.

Dynamic load

The wind that is blowing over a bridge is turbulent. The wind speed is changing in space and time (Figure 2).

In any point of the structure, the wind is changing in time around the average value V_m, with a horizontal turbulent component u and a vertical turbulent component w, as shown in Figure 3. The fluctuation of the vertical component produces a fluctuating angle of attack of the wind $\alpha(t)$ (Figure 3). If the deck, which is the most sensitive part of the bridge to the wind, is considered, the wind speed in one point of the deck changes in intensity V (t) and direction, with an angle of attack $\alpha(t)$ (Figure 3).

As a consequence, with reference to Figure 4, where the aerodynamic coefficients of the Messina bridge are reported as a function of the angle of attack α (positive if nose-up), the lift, moment and drag forces are changing in time, being functions of $V(t)$ and $\alpha(t)$. Similar fluctuations of the aerodynamic forces due to turbulence are produced by the wind on the towers and the cables. The turbulence effect produces, as observed, a fluctuating load on all the bridge components. The motion induced by this load causes fatigue problems on the structure. The turbulence-induced bridge motion, as a consequence, must be confined under limits defined by practice and specifications.

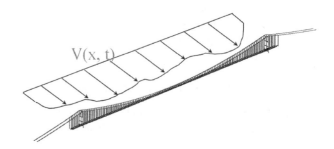

Figure 2. Wind turbulence: space variations of the wind speed along the span in the same time.

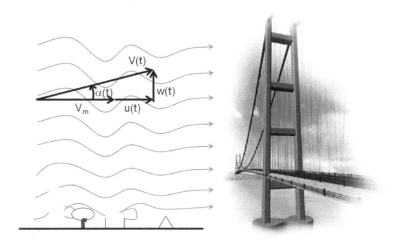

Figure 3. Average and turbulent components of the wind (α defined with reference to V_m direction).

Instability of the bridge

Two forms of instability can occur in a bridge:

- single degree of freedom instability,
- two degrees of freedom coupled flutter instability.

These forms of instability are due to the motion-induced aerodynamic forces acting on the deck. In particular, these forces are functions of the displacement and velocity of the deck. This fact can be easily understood by analysing the simple example in Figure 5, in which the deck is moving in a vertical direction (z) with associated vertical velocity ($\dot{z}$).

In this case, even under a constant wind V_m, the relative velocity of the wind with respect to the deck becomes V_{rel} and has an horizontal component equal to V_m and a vertical component equal to $-\dot{z}$: the resulting wind angle of attack

α (negative according to the sign conventions of Figure 3) is therefore a function of $\dot{z}$. The higher amplitudes of the vertical motion (z) of the deck are in correspondence with the natural frequency of the bridge excited by the incoming turbulence. For each of these components, $\dot{z}$ is harmonically changing and the same occurs for the angle of attack α.

Analysing the $C_D(\alpha)$, $C_L(\alpha)$ and $C_M(\alpha)$ curves shown in Figure 4, it is observed that the coefficients change as a function of α and, as a consequence, the aerodynamic forces applied to the deck will be functions of the vertical velocity $\dot{z}$ of the deck. If these forces dissipate energy, the system is stable; otherwise, if these forces introduce energy into the system, the amplitude increases and the system becomes unstable. The single degree of freedom instability is due to the shape of the deck. Large frontal area of the deck can produce negative slopes of the lift and moment aerodynamic coefficients that are responsible for this type of instability (see Figure 11, Tacoma).

Deck sections with an airfoil-like geometry have positive slope of the lift and moment coefficients and they do not have single degree of freedom instability. Nevertheless, they suffer of two degree of freedom flutter instability, i.e. a coupling between one torsional mode of the deck with the corresponding vertical one. The motion-dependent aerodynamic forces are able to make the two

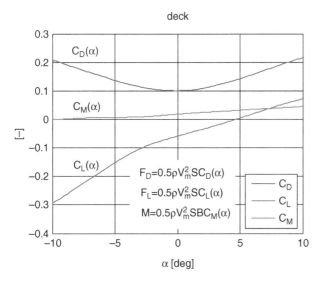

Figure 4. Static aerodynamic coefficients as function of the angle of attack α (positive if nose-up). (S: reference area; B: Deck chord; ρ: air density; V_m: average wind speed).

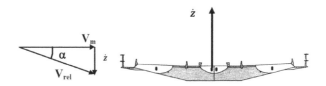

Figure 5. Vertical motion of the deck section and relative wind velocity.

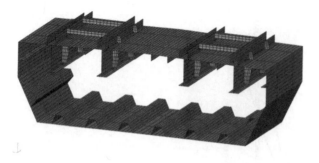

Figure 6.   FEM discretisation with plate elements of the railway box girder of the Messina Bridge.

frequencies come closer and produce this kind of instability.

Vortex shedding

Vortex shedding is produced in a large amount by bluff bodies, like towers, but also by the deck due to the box girder corners, wind barriers or any other device placed on the deck. Vortex shedding produces a fluctuating load at the Strouhal frequency f_S, related to the wind speed and the body dimensions, by the formula: $f_S = c_S V/D$, where c_S is the Strouhal number, V is the wind speed and D is a reference dimension of the body. Furthermore, c_S, for a circular cross section, is around 0.2; for rectangular cross sections – like the tower legs – is between 0.11 and 0.18; for the deck – taking the deck width as a reference dimension – it can be around 1 or less.

When the vortex shedding frequency equals one of the bridge natural frequencies, the vortex shedding induced motion is amplified and its amplitude can reach high values leading to fatigue of the structures or comfort problems. This is one of major concerns in the design of the bridge and is investigated by special wind tunnel tests, as described in the following dedicated paragraph.

Analytical methods for the analysis of the bridge response to the wind

It is common practice to develop finite element schematisations of the bridge (Figure 2) to identify its response to the different actions of interest: wind, earthquake and traffic. With respect to wind action, generally beam elements are used for deck and tower, while taut string or tensioned beam elements are used for the cables. These models are generally used to compute the global behaviour of the bridge to the different actions and they can be nonlinear or linear, as a function of the type of excitation.

A nonlinear approach is used to identify the static position of the bridge under the permanent loads, possibly considering also the static loads due to traffic and the average wind component (Diana, Bruni, & Rocchi, 2005). For what concerns the dynamic load, a linear approach is generally used, being the linearisation performed around the static equilibrium configuration (Chen, Kareem, & Matsumoto, 2000; Diana et al., 2005; Jain, Jones, & Scanlan, 1996a, 1996b; Katsuchi, Jones, & Scanlan, 1999).

To compute the stress induced by the dynamic load – or local effects – a more sophisticated schematisation of the deck and tower is used, employing plate elements, as shown in Figure 6. Once the bridge dynamic response through the simplified equivalent beam model is computed, these results are used as input for the more sophisticated models to perform the stress analysis. With the finite elements model made of beams elements, it is easy to compute the bridge natural frequencies (Figure 7) which are fundamental for the analysis of the bridge response to the wind action. For instance, the ratio between the torsional and vertical frequencies of homologues mode shapes (modes having a high value of Modal Assurance Criterion like those reported in Figure 7) is very important for the flutter instability problem: the greater this ratio, the higher is the flutter velocity, as aforementioned.

In order to reproduce the bridge response to the turbulent wind, the dynamic load due to the incoming turbulence has to be applied to the finite element model (FEM) schematisation. Knowing the structure of the wind in the site surrounding the bridge, it is possible, by special tools (Chen & Kareem, 2001a, 2001b; Ding, Zhu, & Xiang, 2006), to reproduce a space and time wind distribution, as shown in Figure 2. As already mentioned, the forces acting on the bridge are a function of the incoming turbulence. On the other hand, the bridge motion produces aerodynamic forces which are motion dependent. As a consequence, the aerodynamic forces acting on the bridge are functions both of the incoming turbulence and of the bridge motion.

Figure 7.   Vertical and torsional mode shapes.

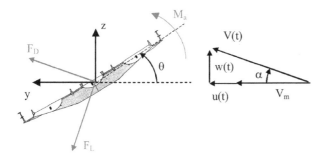

Figure 8.   Aerodynamic forces acting on the deck.

In order to reproduce the over mentioned wind forces in the bridge FEM, the following equation has to be considered:

$$M_s\ddot{X} + R_s\dot{X} + K_sX = F_a(X, \dot{X}, s(t)), \qquad (1)$$

where M_s, R_s, K_s are respectively the mass, damping and stiffness matrices of the bridge structure, linearised in the neighbourhood of the static equilibrium position, X is a vector containing the bridge degrees of freedom and the incoming turbulence vector $s(t)$ contains the incoming wind turbulent velocity in the different sections of the bridge. The X vector defines the motion of the bridge components: deck, towers, cables through the displacement of the i-th section of each single j-th component:

$$X_{i,j} = \begin{Bmatrix} y_{i,j} \\ z_{i,j} \\ \vartheta_{i,j} \end{Bmatrix}.$$

where F_a is the vector defining the aerodynamic forces acting on the different sections of the bridge deck, towers, cables; F_a is function, as previously said, of the bridge motion X and of the associated velocity $\dot{X}$, as well as of the incoming wind turbulent velocity components, in the different sections of the bridge, where F_a is applied.

The identification of this nonlinear function through wind tunnel tests is a complex matter and it is, for sure, one of the most advanced topics of the research in this area (Diana, Resta, & Rocchi, 2008; Diana, Rocchi, Argentini, & Muggiasca, 2010; Wu & Kareem, 2011). The most widely used approach consists in linearising the aerodynamic forces in the neighbourhood of the static equilibrium position of the bridge X_0, computed under a given average wind V_m. This leads to:

$$M_s\ddot{X} + R_s\dot{X} + K_s(X - X_0) = F_{a0}\left(X_0, 0, V_{m0}\right)$$
$$+ \left(\frac{\partial F_a}{\partial X}\right)_{X_0,0,V_{m0}} (X - X_0) + \left(\frac{\partial F_a}{\partial \dot{X}}\right)_{X_0,0,V_{m0}}$$
$$\dot{X} + \left(\frac{\partial F_a}{\partial s}\right)_{X_0,0,V_{m0}} (s(t) - V_{m0}), \qquad (2)$$

The aerodynamic forces F_a must be identified through suitable wind tunnel tests on the deck, towers and cables,

as will be better explained in the next paragraphs. It is very important to well identify the aerodynamic forces acting on the deck because these forces play the most important role in the bridge motion and stability. The aerodynamic forces acting on a deck section (Figure 8) are, as already noticed, function of the vertical (z), horizontal (y) and torsional (θ) deck motion, as well as of the incoming wind turbulent velocity (u, w).

Typically, the aerodynamic forces due to the incoming turbulence are defined by a lift component F_L, normal to the direction of the wind $V(t)$, and a drag component F_D parallel to $V(t)$, both applied at the centre of mass of the section, plus a moment of the aerodynamic forces M_a, around the axis passing through the centre of mass, as presented in Figure 8. Coming back to Equation (1), the lift, drag and moment on each section of the deck represent a sub-vector of the vector containing all the lift, drag and moment forces applied to all the bridge sections while X is the vector of all the section displacements and $\dot{X}$ of all the section velocities. If we use a linearised approach as in Equation (1) with the notation:

$$\left(\frac{\partial F_a}{\partial X}\right)_{X_0,0,V_{m0}} = -K_a; \quad \left(\frac{\partial F_a}{\partial \dot{X}}\right)_{X_0,0,V_{m0}} = -R_a;$$
$$\left(\frac{\partial F_a}{\partial s}\right)_{X_0,0,V_{m0}} = A_m, \qquad (3)$$

then Equation (1) can be written as follows:

$$M_s\ddot{X} + (R_s + R_a)\dot{X} + (K_s + K_a)X = A_m(s(t) - V_{m0}), \quad (4)$$

where $s(t) - V_{m0}$ is a vector defined by the w, u and v components of the wind turbulence in the different points of the structure; R_a and K_a are the equivalent damping and stiffness matrices due to the aerodynamic forces.

Hence, the linearisation of the aerodynamic forces produces an equivalent elastic and damping system that is different from the structural one, changing its natural frequencies and the overall damping: an aeroelastic system is produced. The K_a matrix, as well as the R_a matrix, is not symmetric and, as already mentioned, it can give rise to flutter instability. Furthermore, R_a, if not positive defined, can give rise to the so-called one degree of freedom instability according to a torsional or vertical motion, when the slope of the lift or moment coefficients is not positive, while A_m is the so-called 'Admittance matrix'.

All the parameters required to define the K_a, R_a and A_m matrices must be identified through wind tunnel tests in order to reproduce the bridge response to the turbulent wind. Equation (4) can also be solved in the frequency domain (Jain et al., 1996a, 1996b; Katsuchi et al., 1999; Minh, Miyata, Yamada, & Sanada, 1999; Minh, Yamada, Miyata, & Katsuchi, 2000); in this case, the wind input is the spectrum of the incoming turbulence.

Figure 9. Deck sectional model in wind tunnel (external balance).

Wind tunnel tests

This section is aimed at the description of the types of wind tunnel tests performed in order to verify the bridge performances to the wind action. The procedures, adopted for the most important long bridges at the design or construction stage, will be described. These represent the state of the art on the subject. The tests on deck, towers and cables will be separately considered because the adopted approach is, even if in a small amount, different.

Deck

The most important wind tunnel test for the deck, as well as for the other components, is the definition of the static aerodynamic coefficients as a function of the angle of attack. This test is made in a wind tunnel on a sectional model of the deck (Figure 9) using a dynamometric system, generally placed outside the wind tunnel test section (Cigada, Falco, & Zasso, 2001; Niu, Chen, Liu, Han, &

Hua, 2011). The model can rotate around a deck longitudinal axis and the lift, drag and moment aerodynamic forces are measured as a function of the wind angle of attack, by changing the deck angle of rotation.

Some laboratories use a deck sectional model with a dynamometric central part (Figure 10). This type of device is also very useful to measure the dynamic aerodynamic forces, as it will be better explained in the sequence. The output of this type of tests is shown in Figure 11 for different deck types (taken from Brancaleoni et al. (2009)).

The static forces on all the bridge components, due to the average wind, can be derived from Figure 11 and similar ones for towers and cables. The static forces on all the bridge, due to the average wind, can be computed through an FEM simulation in order to define the deflection of the bridge and the related stresses. To compute the aerodynamic forces as a function of deck motion and incoming turbulence, as reported in the previous paragraph, the lift, drag and moment forces have

Figure 10. Deck sectional model in wind tunnel (internal balance).

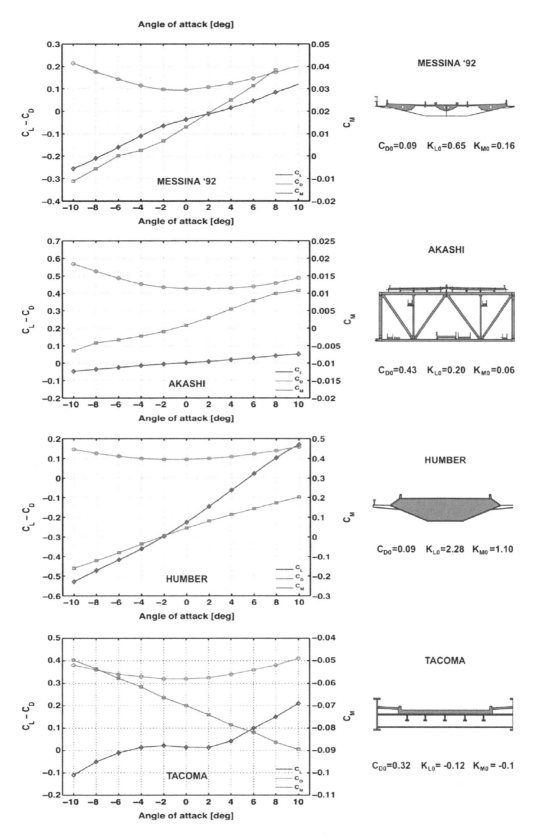

Figure 11.   Aerodynamic coefficients of different deck geometry (Brancaleoni et al., 2009).

Figure 12. Sectional model on oil dynamic actuators driving its rigid vertical and torsional motion.

to identified as a function of the y, z and θ coordinates representing the deck motion and of the w an u velocity components defining the incoming wind turbulence. Regarding the forces identification as a function of the bridge motion, two methods are available: the free motion method and the forced method.

Free motion method

During free motion tests, the deck sectional model is let free to vibrate under the mean wind speed action, by means of an elastic supporting system that should reproduce in the right scale the first flexural and torsional natural modes of the bridge. The response of the system to imposed initial condition is measured and flutter derivatives coefficients are defined by computing the variation of the structural response compared to the values obtained in still air (Bartoli, Contri, Mannini, & Righi, 2009; Chowdhury & Sarkar 2003; Gu, Zhang, & Xiang, 2000; Ibrahim & Mikulcik, 1977; Scanlan & Tomko, 1971).

Forced motion method

This method is more expensive than the previous because it requires specific wind tunnel test rigs to control the deck model motion, and the adoption of specific models that allow for the measurement of the aerodynamic forces taking into account the inertial effects. The major drawback is that this method is much more reliable than the previous one because it allows more repeatable and controlled laboratory test conditions. A deck sectional model is forced to vibrate along the y, z and θ directions and the lift, drag and moment forces are measured for a given constant velocity V_m (Figure 12).

As already mentioned, the forces are nonlinear functions of the motion. In more details, for the deck (see Figure 8):

$$\text{Drag} = F_D = \frac{1}{2}\rho V_m^2 S C_D(X, \dot{X}),$$

$$\text{Lift} = F_L = \frac{1}{2}\rho V_m^2 S C_L(X, \dot{X}),\tag{5}$$

$$\text{Moment} = F_M = \frac{1}{2}\rho V_m^2 S B C_M(X, \dot{X}),$$

while:

$$X = \begin{Bmatrix} y \\ z \\ \vartheta \end{Bmatrix},$$

and ρ is the air density and S and B are respectively the reference area and the deck chord. The identification of the C_D, C_L and C_M coefficients, nonlinear functions of $X, \dot{X}$, as already noted, is not an easy task and represents a research topic.

Various methods have been developed by some researchers: rheological models are used in (Diana et al., 2008, 2010) neural network algorithms are used in (Wu & Kareem, 2011). What is generally done is to linearise the forces F_a of Equation (1) around a static equilibrium position of the deck identified by X_0:

$$F = \begin{Bmatrix} F_D \\ F_L \\ F_M \end{Bmatrix} = \frac{1}{2}\rho V^2 S\big([K_a]\,X + [R_a]\,\dot{X}\big).\tag{6}$$

The 3×3 K_a and R_a matrices contain the flutter derivatives coefficients according to the definition of the

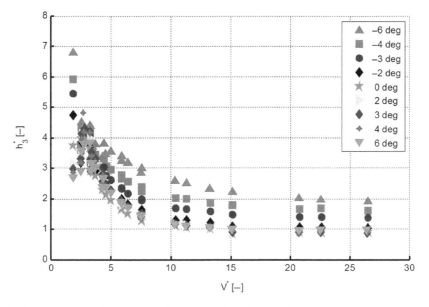

Figure 13. Flutter derivatives (Messina bridge; angle of attack positive if nose-up).

vectors F and X (see Equations (6) and (5), respectively):

$$[R_a] = \frac{1}{2}\rho V_m^2 S \begin{bmatrix} -p_5^* \frac{1}{V_m} & -p_1^* \frac{1}{V_m} & -p_2^* \frac{1}{V_m} \\ -h_5^* \frac{1}{V_m} & -h_1^* \frac{1}{V_m} & -h_2^* \frac{1}{V_m} \\ -a_5^* \frac{B}{V_m} & -a_1^* \frac{B}{V_m} & -a_2^* \frac{B}{V_m} \end{bmatrix}. \quad (7)$$

$$[K_a] = \frac{1}{2}\rho V_m^2 S \begin{bmatrix} p_6^* \frac{\pi}{2V_\omega^{*2}} \frac{1}{B} & p_4^* \frac{\pi}{2V_\omega^{*2}} \frac{1}{B} & p_3^* \\ h_6^* \frac{\pi}{2V_\omega^{*2}} \frac{1}{B} & h_4^* \frac{\pi}{2V_\omega^{*2}} \frac{1}{B} & -h_3^* \\ a_6^* \frac{\pi}{2V_\omega^{*2}} & a_4^* \frac{\pi}{2V_\omega^{*2}} & -a_3^* B \end{bmatrix}. \quad (8)$$

Actually, the identification of the flutter derivatives is not an easy task. The best results are obtained, imposing a harmonic motion to the deck sectional model and measuring the aerodynamic forces on the central part. From these tests, the flutter derivatives value as a function of the frequency of the motion, as well as of the static equilibrium position angle θ_0 around which the tests are done, can be derived.

The frequency is generally reported in terms of reduced velocity:

$$V^* = \frac{V_m}{fB}, \quad (9)$$

where B is the deck width, f is the frequency of the imposed motion and V_m is the velocity at which the test is performed. As an example, the h_3^* value of Equation (8) for the Messina deck are reported in Figure 13: the reduced velocity is reported in the abscissa and the different curves are related to different angles of attack θ_0 around which the harmonic motion is applied.

Identification of the aerodynamic admittance matrix

A method used in the Politecnico di Milano wind tunnel and also by other researchers (Niu et al., 2011), to identify the aerodynamic forces as a function of the incoming turbulence, is to generate a highly correlated wind turbulence along the deck sectional model by an active device of the type shown in the Figure 14.

The airfoils can rotate around a fixed axis with a given motion, which is generally harmonic, with different frequencies. This generates a given fluctuation of the incoming wind, controlled in amplitude, that can be measured through suitable anemometers, like the hot-wire or multi hole pressure probes anemometers. The aerodynamic forces are measured on a sectional deck model for a given average angle of attack θ_0 and the coefficients of the admittance matrix are identified, in a similar way as the flutter derivatives.

The aerodynamic admittances are functions of θ_0 and of the reduced velocity at which the test is performed. As an example, the admittance function coefficients are reported in Figure 15. They are related to the following expressions of the aerodynamic forces:

$$F = \frac{1}{2}\rho V_m^2 S[\chi_a]w. \quad (10)$$

$$F_y = \frac{1}{2}\rho V_m^2 S\left[\mathrm{Re}\left(\chi_{y,w}\frac{w}{V}\right) + \mathrm{Im}\left(\chi_{y,w}\frac{w}{V}\right)\right],$$

$$F_z = \frac{1}{2}\rho V_m^2 S\left[\mathrm{Re}\left(\chi_{z,w}\frac{w}{V}\right) + \mathrm{Im}\left(\chi_{z,w}\frac{w}{V}\right)\right], \quad (11)$$

$$F_\vartheta = \frac{1}{2}\rho V_m^2 SL\left[\mathrm{Re}\left(\chi_{\vartheta,w}\frac{w}{V}\right) + \mathrm{Im}\left(\chi_{\vartheta,w}\frac{w}{V}\right)\right].$$

Once the flutter derivatives and the aerodynamic admittance functions are known, it is possible to compute the bridge response to the turbulent wind. First, the static

Figure 14. Active turbulence generator.

equilibrium position of the bridge under the turbulent wind must be computed as already explained. For each bridge section, a time history of the turbulent wind must be generated. The spectrum of the single time history can be identified and Equation (4) can be solved, for instance, in the frequency domain, taking into account that the flutter derivatives and the admittance function are frequency dependent.

Other methods are available to solve this problem in the time domain (Caracoglia & Jones, 2003), generally with modal approach (Diana et al., 2005), and some of these methods, as already mentioned, take also into account the nonlinear effects of the aerodynamic forces (Chen & Kareem, 2001a, 2001b; Diana et al., 1995). Some examples of output results are presented in the sequence. Figure 16 reports the static deflection of the Messina bridge deck under the action of mean wind forces for different mean wind speeds. Figure 17 reports RMS values of the vertical acceleration of the different deck section along the bridge axis of the Messina bridge deck under the action of turbulent wind ($I_u(z = 70\,\mathrm{m}) = 13.8\%$; $I_w(z = 70\,\mathrm{m}) = 6.9\%$).

The standard deviation of the normal stresses, induced by the buffeting and aeroelastic response, is computed on the downwind road box ($y > 0$) considering the point A reported in Figure 18. Figure 19 illustrates the standard deviation of the stress in the road boxes along the bridge axis for a wind speed of 30 m/s and $I_u = 13.8\%$. The same quantity is reported, in Figure 20, for the point B located on the railway box.

Vortex-induced vibrations

Deck

Vortex-induced vibrations (VIVs) represent one of the most important aspects of the deck shape design. Even though the bridge deck of the last generation has an airfoil-like geometry, flow separation occurs in correspondence of with deck sharp corners, in presence of adverse pressure gradients and of wind shields that are usually adopted to protect traffic. VIV is a serious problem that took also a part in the collapse of the Tacoma Narrow Bridge (Billah & Scanlan, 1990) and has to be carefully analysed. In fact,

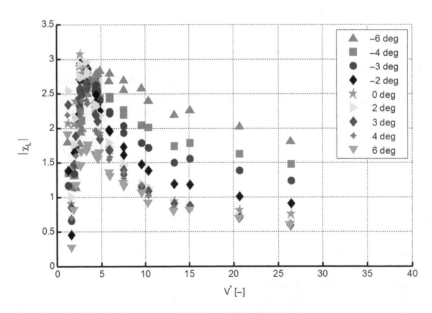

Figure 15. Admittance function (Messina bridge; angle of attack positive if nose-up).

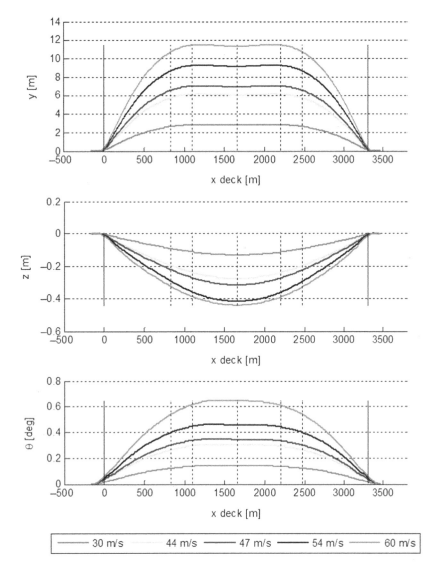

Figure 16.   Static deformation under different mean wind speeds.

even if it can be separated from the stability problem for the modern bridges, it can lead the deck to reach very large vibration amplitudes, also at very low wind speeds having a high probability of occurrence, in case of low Scruton numbers:

$$Sc = 2\pi\frac{m\xi_s}{\rho B^2},\qquad(12)$$

where m is the linear mass, B is the deck chord, ρ is the air density and ξ_s is the structural damping ratio.

Nevertheless, the problem is widely known and studied and past and present experiences highlighted that it represents one of the major concerns in the bridge design. In 2010, Russian authorities shut down the Volga River bridge because of the VIV. For the Storebealt bridge, guide vanes were installed close to the bottom plate/lower side panel joints because mitigation of the vortex-induced oscillations occurred during the final phases of girder erection. The solution was identified after an ad hoc monitoring activity on the bridge at full scale and a specific wind tunnel activity, and its realisation required a specially designed gantry/working platform.

A similar solution was adopted for the twin deck Stonecutters bridge. In both cases, wind tunnel tests on a large sectional model (1:60 for the Storebealt and 1:20 for the Stonecutters) are required to properly design counter-measures that have to be effective to the high Reynolds numbers of the bridge aerodynamics. Wind tunnel tests are performed on elastically suspended models or on taut string models (Figure 21). Special attention is paid to keep the Scrouton number as low as possible by using light

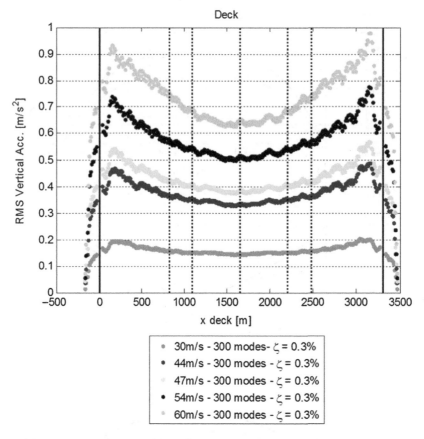

Figure 17.   RMS values of the vertical acceleration of the different deck sections along the bridge axis.

models with low structural damping, in order to better highlight vortex shedding phenomena.

The model response in the lock-in range is usually measured in order to define:

(1) the oscillation amplitudes reached at regime condition throughout the lock-in range;
(2) the amount of energy that the flow is able to insert in the system that may be expressed by a negative value of the aerodynamic damping coefficient measured during build-up tests performed by releasing the deck model under mean wind condition and measuring the growing vibration amplitude starting from rest.

While single-box deck shapes generally show a single mechanism of vortex shedding, basically related to the vortexes detached in the girder wake, multi-box decks may present different vortex shedding phenomena because of

Figure 18.   Considered point on the road and railway boxes of the Messina bridge deck section.

the different possibility of flow interactions between the wake of the upwind box and the other boxes.

The different vortex shedding mechanism may excite both the vertical and the torsional motion of the bridge. As an example, Figure 22 presents the two lock-in ranges for the flexural motion of the Messina Bridge multi-box deck section for different values of Scrouton number. From the maximum values of the amplitudes measured in the lock-in range at different Sc, it is possible to define a figure that reports the non-dimensional amplitudes as a function of Sc.

In Figure 23, the vertical amplitudes of the deck divided by the deck chord B are reported as function of Sc. From this type of figure, it is possible to identify what is the damping of the structure able to control VIVs. Making reference to Figure 23, the three vertical dashed lines represent three different values of Sc that for the Messina bridge corresponding to three different levels of non-dimensional ratios of structural damping equal to: 2×10^{-3}, 3×10^{-3} and 5×10^{-3}. In order to avoid VIVs, the structural damping of the Messina bridge should be greater than 3×10^{-3}.

To avoid VIVs also for lower values of structural damping that are not expected, the solution is to adopt porous screens between the railway box and the road boxes

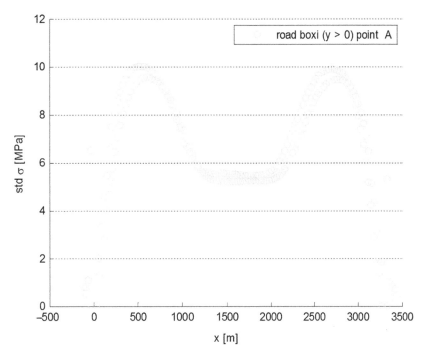

Figure 19.   Standard deviation of the stresses at the point A of the different deck sections along the bridge axis.

(solution MB09 reported in Figure 24). In this case, no VIVs are produced. Measurements of the surface pressure distribution on each single box represent a valuable tool to a better understand of the vortex shedding phenomenon.

In Figure 25, the pressure distribution measured on the Messina bridge model, when it experiences the largest vibrations (lock-in), is reported. The length of the arrows is proportional to the magnitude of the pressure coefficient fluctuations at the vortex shedding frequency:

$$C_P = \frac{p}{(1/2)\rho V_m^2}, \qquad (13)$$

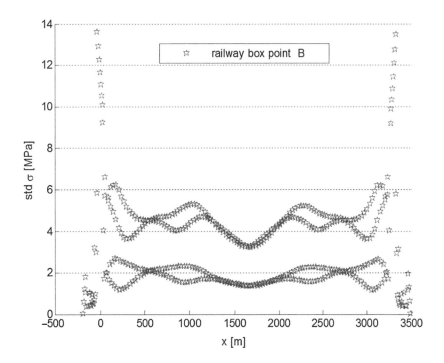

Figure 20.   Standard deviation of the stresses at the point B of the different deck sections along the bridge axis.

Figure 21. Free motion experimental set-up: Suspension system, tensioned steel stays and spring system.

where p is the pressure and V_m is the wind speed. The vortex shedding is mainly related to the capability of the flow to be attached or separated on the curvilinear profile of the upwind box that is run over by the undisturbed flow. The vortex street is generated in the gap between the downwind roadway girder and the railway gird where a strong pressure fluctuation is observable. The aerodynamics of the downwind boxes is, in fact, strongly influenced by the disturbances introduced in the flow by the wind interaction with the upwind box and by the gap between the boxes.

Tower

The main problem that has to be addressed during the aerodynamic design of bridge towers is VIV both if we consider cable-stayed bridges that usually have single leg or reversed Y-shape towers or suspension bridges that have H-shape towers with two legs connected by one or more transverse beams. In both cases, specific wind tunnel tests are performed using sectional or aeroelastic models. The usual wind activities on towers are

(1) The measurement of the static aerodynamic coefficients on tower sectional model.
(2) The measurement of VIV response using elastically suspended tower sectional models.
(3) The measurement of VIV response using aeroelastic models.

Static coefficients are measured in a similar way of what is done on the deck, adopting sectional models with internal or external dynamometers and measuring the aerodynamic forces at different angles of attack. Figure 26 shows a tower sectional model during the wind tunnel tests. For towers with two legs and transversal beams, the aerodynamic coefficients are measured separately on each leg by using a model that represents a tandem arrangement of the two legs where transversal beams are not present.

VIVs of bridge towers are investigated in the wind tunnel by measuring the dynamic response of sectional models that are elastically suspended in the flow in a similar manner of what is already described for the deck or

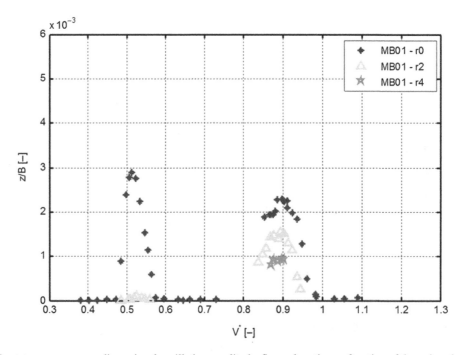

Figure 22. Steady state response: non-dimensional oscillation amplitude, flexural motion as function of the reduced velocity varying the Scrouton number ($r0$: Sc = 0.007; $r2$: Sc = 0.022; $r4$: Sc = 0.033).

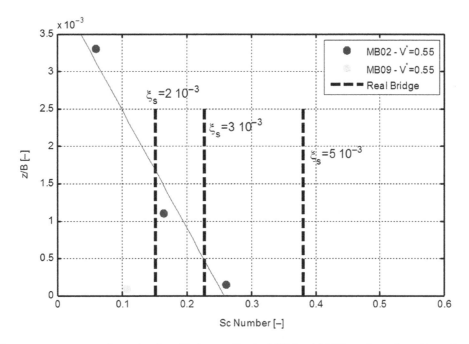

Figure 23. Steady state response: non-dimensional oscillation amplitude, MB02 and MB09 configurations, flexural motion as a function of Scruton Number.

of aeroelastic models (Figure 27). While the vortex shedding phenomenon is basically two-dimensional on the deck, on the contrary for the tower, it is affected by the three-dimensional end effects at the top and by the presence of the transversal beams along the height. Wind tunnel tests performed on aeroelastic models are therefore aimed to represent the full 3D tower geometry and the related aerodynamic effects (Figure 28).

Moreover, aeroelastic models are used to investigate the different construction stages of the tower (Figure 29) because its dynamic characteristics may largely vary from

the free standing condition to the bridge completion condition. If wind tunnel test results highlighted that VIV may reach dangerous values, suitable countermeasures have to be implemented and design on the basis of the experimental results that have also to consider the dynamic effects of all the scaffolding and cranes that are connected to the structure and participate to the dynamic response of the system.

It is not an easy task to optimise the shape of the tower in order to avoid vortex shedding excitation. The final output of the analysis is similar to that of the deck and the

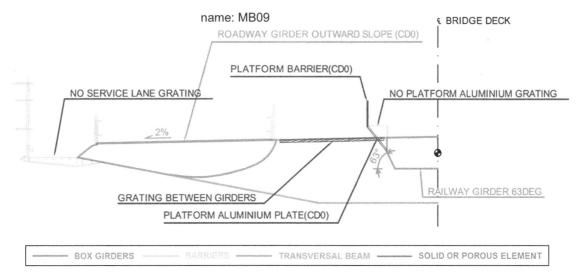

Figure 24. Solution MB09: deck with porous screens between the boxes.

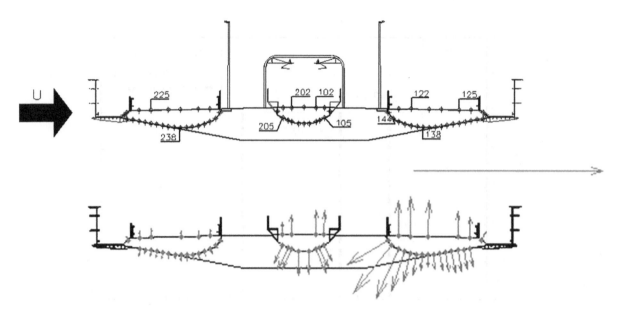

Figure 25. C_p magnitude at $f = 4.21$ Hz ($\alpha = 0°$; $V = 4.64$ m/s).

VIVs are reported as a function of Scrouton number. Tuned mass dampers are generally used to control VIV of the towers.

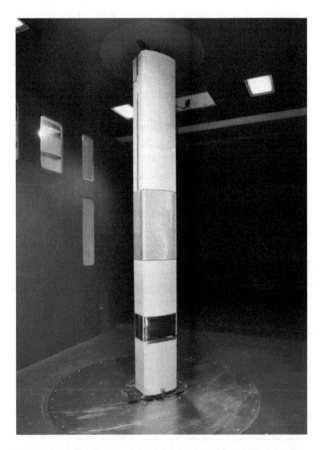

Figure 26.   Cable-stayed bridge tower sectional model in the wind tunnel test section: measure of static aerodynamic coefficients.

Reynolds effect

The scaled reproduction of the wind-structure interaction must necessarily consider Reynolds number effects. To keep the correct ratio between the inertial and the viscous forces, the Reynolds number parameter:

$$\mathrm{Re} = \frac{V_m D}{\nu}, \qquad (14)$$

where D is a reference length and v is the air kinetic viscosity, should be same of the full-scale condition. Due to the practical limitations in the possibility to manage the single parameters, this request is commonly not satisfied, and in the usual wind tunnel practice, a sensitivity analysis on the Reynolds number dependency is performed. Reynolds number dependency is tested by comparing

Figure 27.   Cable-stayed bridge tower sectional model suspended in the wind tunnel test section.

Figure 29. Aeroelastic model of a tower at construction stage.

Figure 28. Free standing tower.

results obtained at different wind speeds and/or on models with different scaling factor.

As an example, in Figure 30, the static aerodynamic coefficients of the Messina bridge deck are reported for different Reynolds numbers. Both the drag coefficient and the aerodynamic moment coefficient show a univocal trend of modification increasing the Reynolds number parameter. In some cases, dealing with curvilinear shapes, it is possible to modify the surface finishing by using an increased roughness in order to simulate similar fluid dynamic effects of higher Reynolds number conditions at lower wind speed (Figure 31).

This wind tunnel practice allows to develop, on a smaller model, similar boundary layer conditions to what is produced at higher Reynolds number on the real structure. Comparing the results of Figure 30 with those reported in Figure 32, obtained at the same wind speed on the same deck sectional model with increased surface roughness, it is possible to observe how the slope of the lift and of the aerodynamic moment coefficient tends to increase at higher Reynolds numbers. This is mainly due to the different fluid dynamic behaviour that is present on the curvilinear part of the upwind box as reported in Figure 33, where the pressure distribution for the nominally smooth and for the rough model are overlapped at the angle of

attack of 8° where Reynolds effects on C_L and C_M are more evident.

Another parameter that has to be controlled during the wind tunnel tests on scaled models, trying to respect the Reynolds number similitude, is the turbulence characteristics of the incoming flow. Different incoming turbulence conditions may lead on the same model to different results in terms of aerodynamic coefficients obtained at the same wind speed. Nominal wind tunnel smooth flow conditions are actually low turbulence conditions that depends on the quality of the flow of the wind tunnel facility ($0.2 < I_u < 2\%$). Although the value of turbulence intensity is very small, in some cases, it may play an important role and it could explain the discrepancies that sometimes arise between results obtained in similar conditions in different wind tunnel plants.

Full aeroelastic models

Full aeroelastic models (Figures 34 and 35) are used to simulate the complex wind structure interaction of the completed or semi completed bridge. Once defined, the aerodynamic behaviour of each single component of the bridge by means of the over described wind tunnel tests the full bridge behaviour may be predicted by numerical approaches relying on numerical models or by a scaled simulation of the full aeroelastic structure. Full aeroelastic models are also used to check if undesired dangerous conditions may arise during the construction stage when the structure has not yet reached its final robustness.

The reliability of the results depends on the quality of the model that has to reproduce in a very small scale the dynamic behaviour of the real structure. There are scaling rules that, if fulfilled, allow to reproduce, in the correct proportions, the aerodynamic and the structural forces in order to be representative of the wind-bridge interaction at full scale. Adopting the Froude scaling rule, the velocity scaling factor λ_v and the frequency scaling factor λ_f are bound to the geometrical scale factor λ_L according to the

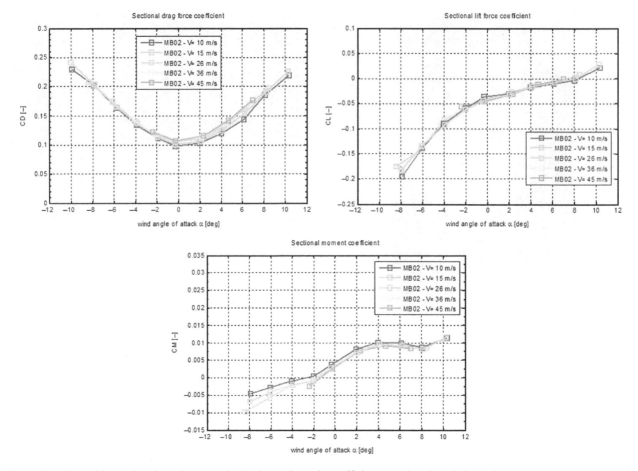

Figure 30. Reynolds number dependency on the deck aerodynamic coefficients varying the wind speed.

following expressions:

$$\lambda_L = \frac{L_{FS}}{L_{Model}} = \qquad (15)$$

$$\lambda_v = \frac{v_{FS}}{v_{Model}} = \sqrt{\lambda_L} \qquad (16)$$

$$\lambda_f = \frac{f_{FS}}{f_{Model}} = \frac{1}{\sqrt{\lambda_L}}, \qquad (17)$$

where FS denotes 'Full Scale'.

The geometrical scale factor being very large, to be able to reproduce the whole, long structure in the wind tunnel test section, a challenging design of the full aeroelastic model is required and a big deal is invested in the model tuning and in the verification of the model structural characteristic with respect to the target values. As an example, in Figure 36, a comparison between a torsional mode shape, experimentally measured on some points along the axis of the aeroelastic full bridge model,

Figure 31. Deck sectional model with increased surface roughness.

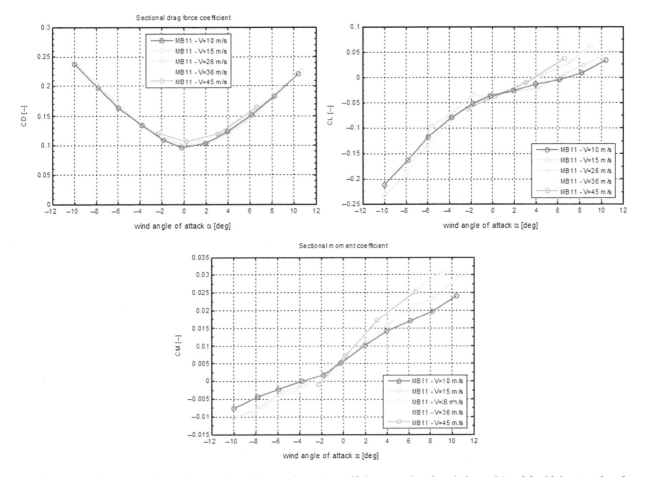

Figure 32. Reynolds number dependency on the deck aerodynamic coefficients varying the wind speed (model with increased surface roughness).

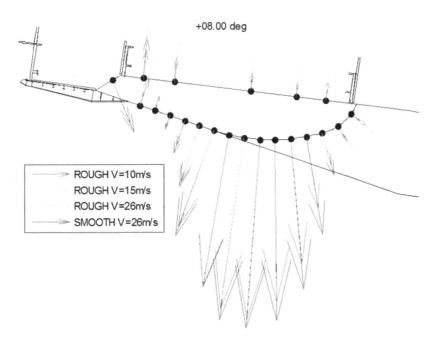

Figure 33. Different boundary layer development (Surface pressure distribution, smooth vs rough model surface, different incoming wing speeds, angle of attack +8°).

Figure 34. Full aeroelastic model on the turning table.

and the target numerical one, obtained by an FEM scheme is reported.

Another key feature is the correct reproduction of the atmospheric boundary layer turbulent characteristics. The goal is reached through the adoption of spires and floor roughness elements that interacting with the wind tunnel flow produce a turbulent mixing of the incoming wind. The turbulent wind characteristics must be scaled in the same

Figure 35. Full aeroelastic model of a cable-stayed bridge.

dimension of the bridge model, in terms of distribution along the bridge axis and in the vertical direction of:

(1) mean wind speed,
(2) turbulence intensities,
(3) power spectra density,
(4) turbulent length scales,
(5) spatial correlation.

Wind tunnel tests on aeroelastic full bridge model are aimed to verify the stability performance of the bridge and the buffeting response. Figure 37 shows how the transfer functions between a frequency sweeping force applied to the aeroelastic model by an electromagnetic shaker and the vertical (on the left) and the torsional (on the right) displacements of the midspan section of the bridge are modified by the aeroelastic effect.

The transfer functions are measured on the model by imposing the force in still air and at two different incoming wind speeds in smooth flow conditions. The modification of the transfer function highlights the effect of the aeroelastic coupling that results in a variation of the natural frequencies of the bridge (with a reduction of the torsional frequency) and in a variation of the slope of the phase diagram in correspondence of the resonance frequency denoting an aeroelastic damping effect.

Figure 38 reports the trend of the non-dimensional damping measured on the first second horizontal, vertical and torsional modes when the bridge model is run over by smooth flow with different mean wind speeds and it is released starting from initial conditions corresponding to the mode shape under analysis. It is possible to observe that for wind speed (reported in model scale) higher than 4 m/s, the second torsional mode assumes a negative value of damping denoting the occurrence of the instability.

From the analysis of the bridge response measured under turbulent wind conditions, it is possible to study the buffeting response of the bridge for different mean wind. Figure 39 plots the RMS value of the vertical acceleration on the different instrumented section of the deck versus the mean wind speed. The trend is parabolic; nevertheless, vortex shedding phenomena occur up to conditions where instability problems arise. An example of the buffeting response to turbulent wind is reported in Figure 40 in terms of the frequency spectrum of the vertical acceleration of the deck section positioned close to midspan, where the contribution of the different structural mode of the structure is evident.

Orography

When the surrounding may influence the flow, a scaled reproduction of the most relevant part of the local orography is considered. Figure 41 depicts an example where a local hill positioned close to the region where a

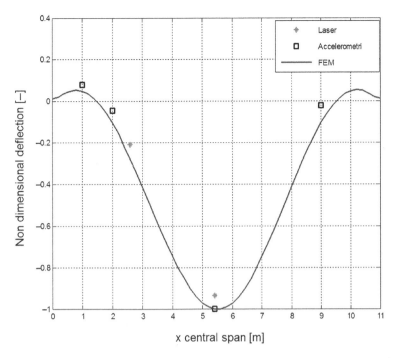

Figure 36. Validation of the aeroelastic model through the comparison between the second torsional mode shape measured in some points along the bridge axis and the FEM target shape for the second torsional mode shape at 0.095 Hz (1.57 Hz model scale, target frequency 1.64).

cable-stayed bridge has to be built may produce deflections and increase the turbulence effects for some wind exposures. In this case, a scaled geometrical reproduction of the land close to the construction zone is visible near the aeroelastic bridge model representing a construction stage.

Conclusions

The common approach to design very long span bridges is to adopt a combined methodology relying on both wind tunnel tests on scaled models and numerical models of the wind-bridge interaction. Wind tunnel tests on sectional

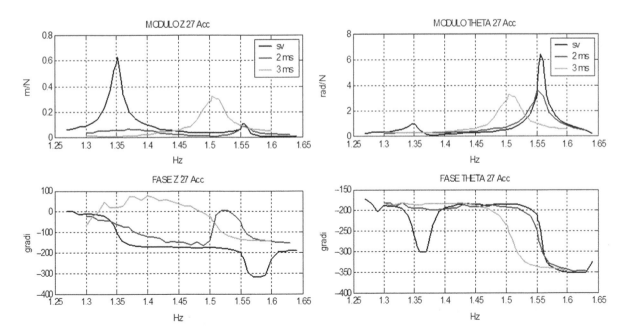

Figure 37. Transfer function at different wind speeds, measured at midspan at 0 m/s; 2 m/s and 3 m/s mean wind speeds, under smooth flow conditions. (Top: Magnitude [m/N or rad/N]; Bottom: Phase [°]).

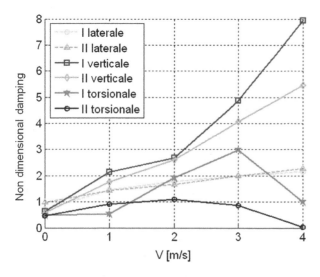

Figure 38. Non-dimensional damping of the first six modes varying the mean wind speed under smooth flow conditions (model scale).

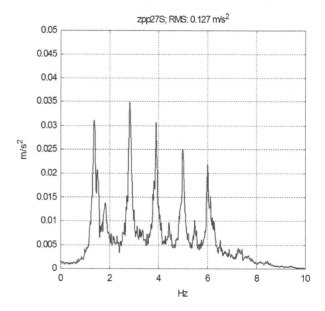

Figure 40. Buffeting response: vertical acceleration measured at mid span.

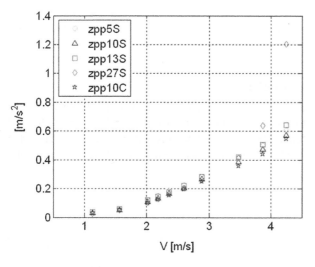

Figure 39. RMS value of the vertical acceleration of section positioned at different positions along the bridge axis varying the mean wind speed under turbulent conditions (5S: 1/11 span south; 10S: 1/5.5 span south; 13S: 1/4 span south; 27S: 1/2; span; 10 C: 1/5.5 span north).

models allow measuring:

- the aerodynamic static coefficients that are used for the definition of the static wind loads and to check if one degree of freedom instability conditions may occur;
- the aerodynamic dynamic coefficients that are used for the definition of the buffeting response and to check if flutter instability conditions may occur;
- define the Scrouton number to limit the VIV of deck and/or tower under a structural safe limit.

Figure 41. Orography reproduction.

Wind tunnel tests and numerical models drive the designer choices, in the early stage of the aerodynamic design process, allowing him to investigate the most important aeroelastic problems: bridge aerodynamic stability and VIV of deck, tower and cables. The most promising solution is therefore analysed in detail through aeroelastic full bridge models in wind tunnel. The results of the full aeroelastic model are compared with the numerical simulations in order to control all the design procedures.

Acknowledgements

Authors want to express their acknowledgements to the staff of the wind engineering group of the Mechanical Department of the Politecnico di Milano. Special acknowledgments are also to

Società Stretto di Messina, who promoted relevant investigations on bridge aeroelasticity.

Notes

1. Email: Giorgio.diana@polimi.it
2. Email: Marco.belloli@polimi.it

References

Bartoli, G., Contri, S., Mannini, C., & Righi, M. (2009). Toward an improvement in the identification of bridge deck flutter derivatives. *Journal of Engineering Mechanics, 135,* 771–785.

Billah, K., & Scanlan, R.H. (1990). Resonance, tacoma narrows bridge failure, and undergraduate physics textbooks. *American Journal of Physics, 59,* 118–124.

Brancaleoni, F., Diana, G., Faccioli, E., Fiammenghi, G., Firth, P.T., Gimsing, N.J., Jamiolkowski, M., Sluszka, P., Solari, G., Valensise, G., & Vullo, E. (2009). *The Messina strait bridge.* Leiden, The Netherlands: CRC Press.

Caracoglia, L., & Jones, N.P. (2003). Time domain vs. frequency domain characterization of aeroelastic forces for bridge deck sections. *Journal of Wind Engineering and Industrial Aerodynamics, 91,* 371–402.

Chen, X., & Kareem, A. (2001a). Aeroelastic analysis of bridges under multicorrelated winds: Integrated state-space approach. *Journal of Engineering Mechanics, 127,* 1124–1134.

Chen, X., & Kareem, A. (2001b). Nonlinear response analysis of long-span bridges under turbulent winds. *Journal of Wind Engineering and Industrial Aerodynamics, 89,* 1335–1350.

Chen, X., Kareem, A., & Matsumoto, M. (2000). Coupled flutter and buffeting response of bridges. *Journal of Engineering Mechanics, 126,* 17–26.

Chowdhury, A.G., & Sarkar, P.P. (2003). A new technique for identification of eighteen flutter derivatives using a three-degree-of freedom section model. *Engineering Structures, 25,* 1763–1772.

Cigada, A., Falco, M., & Zasso, A. (2001). Development of new systems to measure the aerodynamic forces on section models in wind tunnel testing. *Journal of Wind Engineering and Industrial Aerodynamics, 89,* 725–746.

Diana, G., Bruni, S., & Rocchi, D. (2005). A numerical and experimental investigation on aerodynamic non linearities in bridge response to turbulent wind. *Proceedings of the EACWE4.* Prague, Czech Republic, July 11–15, CD-ROM.

Diana, G., Falco, M., Bruni, S., Cigada, A., Larose, G.L., Darnsgaad, A., & Collina, A. (1995). Comparisons between wind tunnel tests on a full aeroelastic model of the proposed bridge over Stretto di Messina and numerical results. *Journal of Wind Engineering and Industrial Aerodynamics, 54–55,* 101–113.

Diana, G., Resta, F., & Rocchi, D. (2008). A new numerical approach to reproduce bridge aerodynamic non-linearities in time domain. *Journal of Wind Engineering and Industrial Aerodynamics, 96,* 1871–1884.

Diana, G., Rocchi, D., Argentini, T., & Muggiasca, S. (2010). Aerodynamic instability of a bridge deck section model: Linear and nonlinear approach to force modeling. *Journal of Wind Engineering and Industrial Aerodynamics, 98,* 363–374.

Ding, Q., Zhu, L., & Xiang, H. (2006). Simulation of stationary Gaussian stochastic wind velocity field. *Wind and Structures, 9,* 231–243.

Gu, M., Zhang, R., & Xiang, H. (2000). Identification of flutter derivativesof bridge decks. *Journal of Wind Engineering and Industrial Aerodynamics, 842,* 151–162.

Ibrahim, S.R., & Mikulcik, E.C. (1977). A method for the direct identification of vibration parameters from the free response. *Shock and Vibration Bulletin, 474,* 183–198.

Jain, A., Jones, N.P., & Scanlan, R.H. (1996a). Coupled aeroelastic and aerodynamic response analysis of long-span bridges. *Journal of Wind Engineering and Industrial Aerodynamics, 60,* 69–80.

Jain, A., Jones, N., & Scanlan, R.H. (1996b). Coupled flutter and buffeting analysis of long-span bridges. *Journal of Structural Engineering, 122,* 716–725.

Katsuchi, H., Jones, N.P., & Scanlan, R.H. (1999). Multimode coupled flutter and buffeting analysis of the Akashi-Kaikyo bridge. *Journal of Structural Engineering, 125,* 60–70.

Minh, N.N., Miyata, T., Yamada, H., & Sanada, Y. (1999). Numerical simulation of wind turbulence and buffeting analysis of long-span bridges. *Journal of Wind Engineering and Industrial Aerodynamics, 83,* 301–315.

Minh, N.N., Yamada, H., Miyata, T., & Katsuchi, H. (2000). Aeroelastic complex mode analysis for coupled gust response of the Akashi Kaikyo bridge model. *Journal of Wind Engineering and Industrial Aerodynamics, 88,* 307–324.

Niu, H.W., Chen, Z.Q., Liu, M.G., Han, Y., & Hua, X.G. (2011). Development of the 3-dof forced vibration device to measure the aerodynamic forces on section models. *Proceedings of ICWE.* Amsterdam, Netherlands, July 10–15, CD-ROM.

Scanlan, R.H., & Tomko, J.J. (1971). Airfoil and bridge deck flutter derivatives. *Journal of Engineering Mechanics Division, 976,* 1717–1737.

Wu, T., & Kareem, A. (2011). Modeling hysteretic nonlinear behavior of bridge aerodynamics via cellular automata nested neural network. *Journal of Wind Engineering and Industrial Aerodynamics, 99,* 378–388.

Operational deformations in long-span bridges

James M.W. Brownjohn, Ki-Young Koo, Andrew Scullion and David List

Long-span bridges deform quasi-statically and dynamically under a range of operational conditions including wind, traffic and thermal loads, in varying patterns, at different timescales and with different amplitudes. While external loads and internal forces can only rarely be measured, there are well-developed technologies for measuring deformations and their time and space derivatives. Performance data can be checked against design limits and used for validating conceptual and numerical models which can in turn be used to estimate the external loads and internal forces. Changes in performance patterns and load–response relationships can also be used directly as a diagnostic tool, but excessive deformations themselves are also a concern in terms of serviceability. This paper describes application of a range of measurement technologies, focusing on response to extreme loads, for suspension bridges over the River Tamar (with 335 m main span) and Humber (with 1410 m man span). The effects of vehicular, thermal and wind loads on these very different structures are compared, showing that apart from rare extreme traffic and wind loads, temporal and spatial temperature variations dominate quasi-static response. Observations of deformation data and sensor performance for the two bridges are used to highlight limitations and redundancies in the instrumentation.

1. Relevance of dynamic and static bridge deformation measurements

Deformation measurements play a critical role in structural identification of bridges (Catbas, Correa-Kijewski, & Aktan, 2013) through both short measurement campaigns and long-term monitoring. In the form of direct measurements of vertical deflection during a vehicle load test (Calçada, Cunha, & Delgado, 2005), deformation data are used to calibrate design and demonstrate fitness for purpose. In the form of accelerations recorded during modal testing (Brownjohn, Magalhães, Caetano, & Cunha, 2010; Kim et al., 2007; Pakzad, Fenves, Kim, & Culler, 2008; Ren, Harik, Blandford, Lenett, & Baseheart, 2004), they provide a direct view of the stiffness and mass properties and their distributions.

These two types of short-term measurement campaigns (load test and modal survey) together with standard assessment tools provide a wealth of information about the state of a structure. All these elements were deployed in a 2010 exercise on US202 bridge 1618-150 in New Jersey (Aktan, 2011) where a range of deformation measurement technologies were demonstrated. For example, load tests typically employ close-range linear variable displacement transformers

(LVDTs) or potentiometers when a fixed reference below the bridge is available, and more exotic non-contacting sensors such as Global Positioning System (GPS) when there is no such fixed base, the usual case for long-span bridges.

For longer-term measurements as part of a bridge structural health monitoring (SHM) system (Ko & Ni, 2005), fewer accelerometers can be deployed, particularly where a modal survey has provided a set of vibration modes (Brownjohn et al., 2010) that can be used as a basis set for mapping a minimal set of measurements to a complete structure. Conversely, identification of the more complex quasi-static three-dimensional deformation patterns requires a more extensive set of instrumentation.

The aim of this paper is to address the particular issue of identifying the nature of such deformations in long-span bridges during normal operation but over long enough periods to capture typical and extreme conditions, providing a useful comparison of their levels and likelihoods. This requires continuous automated capture of deformations and rotations having periods spanning the dynamic range of time scales (where inertia forces are significant) through sub-dynamic (as induced by vehicles and wind gusts) to the daily (weather dependent) and seasonal (climate dependent) range. Along with the wide range of time scales, demanding appropriate frequency response and sample rates, deformations range from microns to meters. Selecting appropriate instrumentation to operate over these time and length scales is a challenge for SHM system designers, with choice depending on the nature of the structure and loading, as well as cost, logistical constraints and purposes of the system.

2. Current viable technologies for measuring bridge deformation

The earliest methods for continuously measuring and monitoring long-span bridge deformations included the use of motion pictures, e.g. to study the aero-elastic problems at Tacoma Narrows Bridge (University of Washington, 1954) and hydrostatic levelling, e.g. to check effects of structural changes to D. Luiz I bridge, Porto (Marecos, 1978). A sophisticated opto-mechanical system was the preferred solution for the deflection measurements of the Tagus River suspension bridge during a remarkably comprehensive exercise of instrumentation and proof load testing in 1966 (Marecos, Castanheta, & Trigo, 1969).

Before widespread use of GPS, opto-electronic technologies were developed for long-span bridges. For example, lasers were used to track vertical and lateral movement in performance of the Foyle Bridge in Northern Ireland (Sloan, Kirkpatrick, Boyd, & Thomson, 1992), and LED-based systems were used in Scandinavian bridges (Myrvoll, Di Biagio, & Hansvold, 1994).

Two radically different technologies were employed in the Humber Bridge monitoring campaign of 1990/1991 (Stephen, Brownjohn, & Taylor, 1993; Zasso, Vergani, Bocciolone, & Evans, 1993). The single axis 'optometer' (Zasso et al., 1993) used for automated monitoring at Humber employed high contrast targets, charge-coupled device arrays and threshold detection for tacking transverse and vertical motion separately. A system based on post-processed pattern recognition and trajectory prediction was used for tracking both axes simultaneously and was subsequently developed for automated real-time tracking of the Second Severn Crossing (Macdonald, Dagless, Thomas, & Taylor, 1997).

There have been experimental GPS applications in the UK, e.g. at Humber (Ashkenazi & Roberts, 1997) and Forth (Roberts, Meng, Brown, & Andrew, 2006) but apparently until 2011 no permanent GPS installation on a bridge. However, GPS has for the last decade been the standard choice for deformation measurements on long-span bridges in the Far East. Examples include the networks of bridges operated by Nanjing's Jiangsu Transportation Institute and Hong Kong Highways Dept. (Wong, 2007), whose designs have been updated for the Forth Replacement Crossing (Kite, Carter, & Hussain, 2010).

Robotic total stations (RTS) are still rare for permanent instrumentation of long-span bridges, exceptions being the

Jiangyin Bridge (Zhou, Ni, & Ko, 2006) and the Tamar Bridge described here. However, some short-term evaluations have been reported (Erdoğan & Gülal, 2011; Stiros & Psimoulis, 2012). GPS is now regarded as a mature technology, alongside extensometers, inclinometers and lasers of various forms. Hydrostatic level sensors (as originally used on Tamar and Tsing Ma bridges) are hardly used these days outside Portugal.

All these instruments have been used for direct measurement of translational deformations; complementary to these, there exist technologies already in use for several decades such as LVDTs and inclinometers for measuring small relative motions between components within a structure. High-quality accelerometers can also be used for quantifying the quasi-static deformations patterns due to slowly varying operational loads that do not engage inertia effects. Good low-noise accelerometers that have capability to measure down to 'DC' or 0 Hz can be used to recover translations in the sensing axis through double integration, but noise imposes a limit on the lowest frequency recoverable. DC units can also measure tilt (inclination), but when translations occur at similar low frequencies as the simultaneous rotations, the two effects may not easily be distinguished (Hjorth-Hansen & Niggard, 1977; Rainer, 1985). Some exemplar permanent deformation monitoring systems are summarised in Table 1; the majority of these have been implemented in the Far East.

3. Monitoring of Tamar and Humber Bridges

With the aim of describing the capabilities and limitations of the technologies mentioned above, deformation monitoring systems on two long-span bridges in the UK are presented. A secondary aim of the paper is to compare and contrast the ways in which these two bridges deform due to operational loads, providing instructive examples of some extreme performance compared with normal ranges, and guidance to inform choice of instrumentation on new bridge projects.

3.1 Deformation monitoring of Tamar Bridge

Tamar Bridge (Figure 1) has provided a useful experience in deformation tracking technologies. Built in 1961 as a conventional suspension bridge with a steel truss and 335 m main span, it provides a link between Plymouth (east, in Devon) and Saltash (west, in Cornwall). To cope with heavier vehicles, the bridge was 'strengthened and widened' in 2000 by radical alterations that included replacing the deck, adding extra lanes cantilevered either side of the truss and around the tower pylons (clearly visible in Figure 1) and adding a set of 18 stay cables to carry the extra load and restore the camber. Two of these cables are installed in inverted U-shaped troughs

Table 1. Example bridge deformation monitoring systems.

Bridge	Span (m)	Built	Installed	Displacement	GPS	Inclinometer
Jiangyin[a]	1385	1998	1998, 2005		8	
Akashi[b]	1991	1998	1998	3 × ext	8	
Nahmae[c]	404	1973	1998	4 × ext		12
Jindo[c]	344	1984	1998			4
Sohae[c]	470	1993		10 × ext, 4 × laser		6
Gwangan[c]	500	1994		3 × ext, 1 × laser		4
Youngjong[c]	300	2000	2001	4 × ext, 3 × laser		10
Tsing Ma[d]	1377	1997	<1997	10 × level, 2 × ext	14	
Ting Kau[d]	475	1998	<1998	2 × ext	7	
Kap Shui Mun[d]	430	1997	<1997	5 × level, 2 × ext	6	
Stonecutters[b]	1018	2009	<2009	34 × ext	20	

[a] Ko and Ni (2005).
[b] Sumitro (2001).
[c] Koh et al. (2009).
[d] Wong (2007).

providing extra stiffening below the lower chords of the truss girder.

Boundary conditions for longitudinal movement at the towers were also altered so that the cantilevers provide axial continuity from the Plymouth side tower (where main cables reach deck level) to the Saltash Tower where expansion joints are installed. To track the effects of the changes during and after the construction, Fugro Ltd installed a structural monitoring system. This is a comprehensive array of sensors including load cells in all 18 additional stays, wind sensors at 3 locations, temperature sensors for air, main cable, deck and truss elements, and a hydrostatic levelling system, amounting to a total of 70 data channels.

3.1.1 Fugro level sensor

This system comprised a fluid-based system with pipes along the main span and level sensing station (LSS) at 1/8

span centres. The system was based on the similar system previously installed by Fugro on bridges in the Lantau Fixed Crossing. The height measurements were specified to be accurate to ±5 mm and while in principle the pressure changes due to level variations should depend only on velocity of pressure waves in the fluid, the acquisition updates every 10 s. The system was overhauled in 2007, but no longer functions.

Figure 2 shows level data for h71, at 3/8 span position from the Plymouth side, for a complete 24-h period (upper plot), then for the rush hour period 4–5 pm (lower plot). The sampling rate is too slow to identify passage of individual heavy vehicles, which at the posted 13.4 m/s speed limit would take 25 s to cross the bridge. While the drifting and jumping of level signals did not allow for reliable long-term tracking over periods of more than a month, it was possible to apply principal component analysis to the seven-component time series. Figure 3 shows the eigenvector corresponding to the first

Figure 1. Tamar Bridge. Left showing additional stays and right showing cantilever and arrangements around Plymouth Tower.

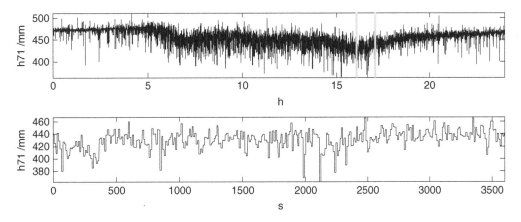

Figure 2. Level sensor data for 1 day and 1 h.

(strongest) principal component of vertical plane deformation. This pattern operates on a daily cycle, primarily driven by temperature, and there is a clear lop-sided pattern resulting from the hybrid structural form and expansion joint arrangement.

3.1.2 Accelerometers and extensometers

In a sequence of upgrades starting in 2006, the Vibration Engineering Section installed various dynamic and static instrumentation components to monitor the deformation behaviour of the bridge. A dynamic data acquisition system (DAQ) was installed in 2006 comprising a set of three Honeywell QA750 servo-accelerometers mounted close to mid-span. Two units (VS and VN) mounted vertically on the north and south sides of the truss measure vertical acceleration with 1 unit (H) measuring horizontal acceleration. They are primarily used for tracking vibration response levels and identifying modal parameters but the H signal can also be used to track rotation of the main span about the long axis.

Since the level sensors and the total station reflectors that superseded them measure only on the south side, accelerometers provide the only means of kinematically connecting the two sides of the bridge. In addition, a set of eight QA650 servo-accelerometers arranged in biaxial pairs was used to study rain-wind induced vibrations of four of the additional stay cables but these have now been removed since the cables have been fitted with dampers. While installed, they provided indirect estimates of the stay cable tensions, useful for corroborating load cell data, and spot checks for the other 12 above-deck stay cables were made by roving a single accelerometer.

Three DAQ channels were also originally used to record extensions at the Saltash Tower expansion joint, using ASM WS12 mechanical (wire) potentiometers (Figure 4). Signals from the complete set of units were available only for a short period and showed that axial separation between Saltash Tower and the main span (north and south sides) closely followed the separation between the spans, i.e. that tower-sidespan motion was negligible. Mechanical disruption to the wire threads (due to birds, workers) caused loss of all signals from March 2007 until 2010 when an electrically decoupled wireless link was established to record the relative motion between tower and main span.

3.1.3 Total positioning system

From the information provided by the hydrostatic levelling system and the Saltash tower extensometers during 2006 and 2007, it was clear that movement of the bridge in the vertical plane, mostly due to thermal expansion, comprised an axial extension of the girder and a lowering of the suspended structure. To study this two-dimensional motion (and to include transverse motion), a total

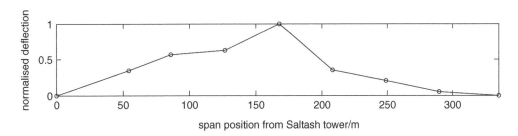

Figure 3. Vector for first principal component of level sensor time series.

Figure 4.   Extensometer (left) and location between Saltash and main spans (right).

positioning system (TPS) was installed, comprising a Leica TCA1201M RTS and a set of 15 reflectors (Figure 5). TPS was chosen following advice from Graz Institute of Engineering Geodesy and Measurement Systems in preference to GPS due to the expected small movements and the lower cost for multiple measurement positions, there being no need for the 'dynamic' sample rates (1 Hz or more) offered by GPS.

The upper portal of the Plymouth Tower provides the best viewpoint for targets on the deck and at reference points, but longitudinal sway prevented a stable vertical plane alignment. The only secure and viable location for the RTS was the roof of the control room adjacent to the roadway, for which the line of sight for distant reflectors around roadway level at the Saltash end results in small angular differences and occasional failure to identify and locate reflectors. This is a secure and convenient location but not optimal in terms of stability and target visibility since with heavy rain, fog and drizzle, location capability is also reduced, resulting in patchy data worse than Figure 6. There are also several practical problems with the RTS unit, including cleaning the casing and servicing the mechanical parts.

The RTS data sampling in time and space is sparse even compared to the level sensors, but the RTS has the advantage of 3D coverage that includes towers. Cross-validation between the two systems during the brief period when both were operational shows an exact linear relationship with unit slope (but considerable scatter).

3.2 Deformation monitoring of Humber Bridge

The Humber Bridge, opened in 1981, has a main span of 1410 m and side-spans of 280 and 530 m aligned almost exactly north–south linking the small towns of Hessle (north) and Barton (south). Aerodynamic steel box sections are 22.5 m wide and 4.5 m deep with 3 m walkway cantilevers each side, and were prefabricated in 18 m long sections with transverse bulkheads at 4.5 m centres. Tapering reinforced concrete towers are 151.5 m high and support twin 0.29 m² main cables with sag of 115.5 m that in turn support the box deck via inclined hangers. Humber was the world's longest span until 1998 but is now surpassed, most notably by Great Belt and Akashi bridges, so that in 2013 it ranks as the seventh longest span.

3.2.1 Previous measurement campaigns

Humber Bridge has provided many opportunities and applications for deformation measurements. The original ambient vibration study (Brownjohn, Dumanoglu, Severn, & Taylor, 1987) and the 1990/1991 extended monitoring (Brownjohn, Curami, Falco, & Zasso, 1994) served the purpose of validating simulation software to be used for analysis and design of other bridges. Specifically, the finite element analysis procedure validated in the 1985 test was used for studying seismic response of the two Bosporus crossings (Dumanoglu, Brownjohn, & Severn, 1992; Dumanoglu & Severn, 1987), while the 1990/1991 study

Figure 5.   RTS unit on control room roof (left) and reflectors on Plymouth side Tower (middle) and Saltash Tower (right).

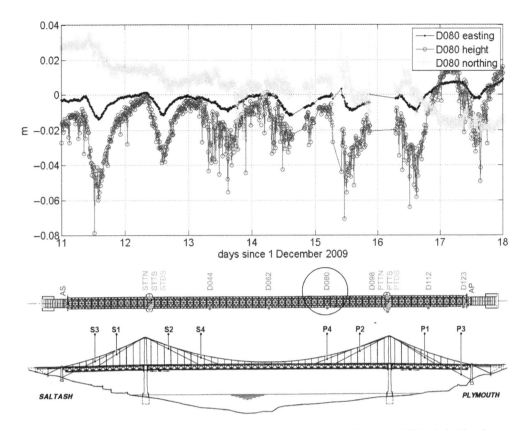

Figure 6. Reflector location plan, additional stay cable identities and deformation at location D080 (circled in plan).

validated procedures for simulating the in-wind performance of the proposed Messina Straits Bridge (Diana, Falco, Cheli, & Cigada, 2003).

The 1990/1991 measurement campaign featuring the use of the two previously mentioned optical tracking systems (Stephen et al., 1993; Zasso et al., 1993) allowed for 2D deflections perpendicular to line of sight to be measured. Combined with mechanical potentiometers measuring motion between the main span and towers and inertial sensors (accelerometers and inclinometers) characterising transverse, vertical and torsional motion, the 1990/1991 study identified relationships between loads and response presented in Table 2. Being operational only during the spring/summer periods of 2 years, the system missed opportunities to capture seasonal effects and some extreme events; moreover, the technology for recovering dynamic (modal) properties left some open questions about lateral oscillations.

Table 2. Humber Bridge deflection–load relationships from 1990 to 1991 monitoring.

Deflection component	Temperature (T/°C)	Wind speed (U/m/s)
Vertical (v/mm)	$v \approx -60T$	$v \approx -0.81U^2$
Rotation (θ/m-rad)		$\theta \approx 0.006U\|U$
Lateral (u/mm)	$u \approx 15T$	$u \approx 1.18U\|U$
Longitudinal (w/mm)	$w \approx 8T$	

3.2.2 Present technology

The present exercise extends and simplifies the capabilities of the earlier system and provides useful management data for Humber Bridge board (HBB), allowing them to make informed decisions on management and intervention. Figure 7 shows the present monitoring system and a sample of sensors. Operational in stages since September 2010, the system comprises the following components:

- GPS base station and two rovers mounted (GMX902 GPS) on the main cables at mid-span plus a Nivel 220 inclinometer inside the box girder at mid-span (operational since May 2011).
- Three QA750 accelerometers mounted inside the steel box girder at mid-span in the same configuration as Tamar (i.e. vertical at east and west extremes, also one horizontal sensor), plus a single RM Young 05305 anemometer mounted on a lamppost.
- Four laser extensometers (Hilti PD4 analog devices converted to digital outputs) were installed at the main span ends at the lower portal beam level adjacent to the bearings, two at each (south and north) tower, with one at each of the east and west ends of the lower tower portal beam (operational from 23 February 2011 to 27 July 2011 then from November 2012). The extensometers were arranged

Figure 7. Humber Bridge (deformation) monitoring system and sensors: GPS antenna on anchorage and main cable; anemometer and accelerometers.

to measure movement of the main span ends away from the towers.

- Thermistors installed at four locations around one 18-m box section (operational from 11 July 2010) supplement signals from HBB. These include additional sensors for air speed, direction and temperature. HBB temperature sensors on the deck surface and structure were also available until October 2012.

As with Tamar Bridge, data fusion is a major concern. Signals from the inclinometer and GPS receivers are recorded at 1 Hz sample rate and managed using Leica GNSS Spider and Spider QC software. Real-time kinematic solutions that apply corrections based on known position of the GPS base station are saved. The Leica software operates on a National Instruments embedded PC that also runs a LabVIEW virtual instrument recording signals from accelerometers and the anemometer at 8 Hz.

Extension and temperature signals are streamed directly to the Internet, so do not rely on the embedded PC; likewise, data feeds from weather and temperature sensors installed by HBB go directly to the remote server for processing. As with Tamar Bridge, automated batch routines process and fuse the data streams into 30-min summary values which are saved to a common time-base and can be accessed using a web-based viewer or a MATLAB interface.

While acceleration data have been recorded almost continuously, there have been data gaps for other deformation sensors. The extensometer interface software has recently been upgraded by HBB and access routes to HBB wind and temperature data have changed, while an extreme wind event destroyed the RM Young anemometer in early 2013. The signal cable to the east GPS antennae was vandalised in late 2012 (and repaired 1 year later), but due to the kinematic relationships already established this was not a serious concern. Despite the practical problems, sufficient data for representative periods have been collected to characterise the bridge deformation mechanisms and the system is maintained to continue providing performance data.

Compared to the 1990/1991 optical deformation system, GPS is a more reliable and practical system for positional tracking, using four extensometers has provided valuable information about the horizontal plane movement of the main span and more comprehensive data are available about structural temperature. There has also been a greater focus on rotations about all axes, and with more sophisticated modal analysis procedures, it has been possible to examine variations in and causality of modal parameter variations with greater confidence.

4. Response ranges and correlations

4.1 Tamar Bridge

For Tamar Bridge, wind effects are apparently relatively insignificant, so Figure 8 gives an impression of deformation ranges and effects of structural temperature variation. Temperature increase directly drives axial extension (which correlates well with deck temperature) and main cable extension (which of course correlates well with main cable temperature). The relationships are not quite linear, particularly for temperatures above (approximately) 15°C on exposed parts of the structure where clear

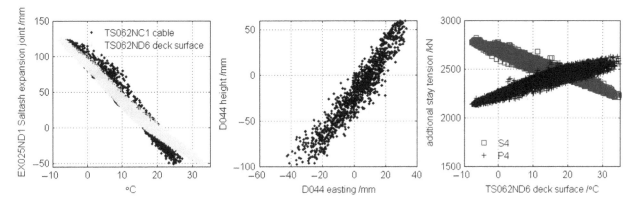

Figure 8.   Tamar Bridge temperature effects on quasi-static deformation (see Figure 6 for stay cable locations).

skies in summer result in large temperature variations across a section.

For increasing temperature, increasing tensions for stay cables P2 and P4 attached to the Plymouth Tower contrast with reducing tensions for stays S2 and S4 attached to the Saltash Tower as main span extension brings it closer. This effect contributes to the lop-sided main span deformation pattern seen in Figure 3, and the breadths of the two groups of data points indicate hysteretic effects due to small time delays in the relationships.

4.1.1 Humber Bridge

Some interesting relationships between response quantities and environmental loads are summarised in Figure 9 for wind and in Figure 10 for temperature. During 2 years of monitoring, winds have caused the largest and most interesting deformations, with 30-min average speeds up to 28 m/s and instantaneous values exceeding 39 m/s (hurricane force). Temperatures have ranged from −8 to 49°C for deck surfacing temperatures and from −6.5 to 30.3°C for air temperatures, all greater than ranges experienced at Tamar.

Lateral (sway) deformation and main span rotation Figure 9 show that it moves away from the wind (negative normal wind is from the west) and dips on the leading edge. In fact, wind also induces downwards deck displacement due to the inverted aerofoil shape, but such displacements are generally smaller than those due to either temperature or vehicle effects, only being noticeable during the strongest winds. There is also a small axial movement linked to lateral sway. In fact, during strong winds, it appears that the whole suspended structure shifts north in westerly winds or south in easterly winds.

Temperature effects are also very clear (Figure 10). Main span vertical displacement and expansion, appearing as reduction of extensometer signals, follow temperature, with slope depending on which temperature reading is used. Unlike Tamar Bridge, no main cable temperature data are available and 'surface' temperature is for a sensor buried in the deck surfacing, while 'box top' is few centimeters below on the inside top surface of the steel box. The internal temperature data provide the cleanest correlation but were not available for much of the time that all four extensometers were operational. Hence, the last (right) plot uses surface temperature to compare axial motion of the deck for bearings at each end along with the GPS northings at mid-span. Extension is more or less even between the two ends, and the GPS shows only a moderate northerly shift at mid-span with temperature increase. Along with the wind effect on northings, this suggests mechanisms relating to the asymmetry of the span configuration.

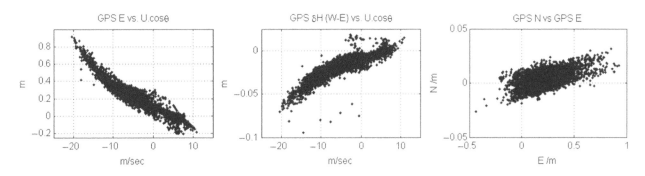

Figure 9.   Humber Bridge: wind effects on quasi-static rotation and horizontal plane deformation. Positive wind is from the west.

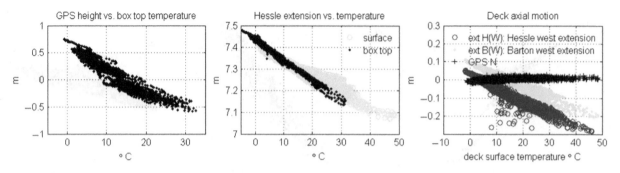

Figure 10. Humber Bridge: temperature effects on quasi-static deck vertical and axial movement.

5. Dynamic displacements

While accelerations may often be dominated by high frequency vibrations excited as a result vehicles reacting to uneven pavement, these translate to insignificant displacements and the greatest contributions are at the fundamental mode frequencies of each type: lateral (L), vertical (V) and torsional (T), and which for these bridges the mode shapes are either symmetric (S) or approximately symmetric about mid-span.

Standard deviation or de-trended root mean square (RMS) of modal displacement is taken to represent the strength of a response in a vibration mode and is derived from RMS of Fourier acceleration amplitudes close to the modal frequency divided by the relevant squared circular frequency. Figure 11 shows the distribution (probability density function) of 30-min RMS displacements for the fundamental modes of each type in each bridge.

The dynamic behaviour of the two bridges is radically different and impacts on the ranges of displacements experienced in vibration modes. Mode frequencies (given in the figure) are much lower for Humber and the displacements are by far the larger, the largest RMS value having been obtained for lateral direction for an extreme wind event. Tamar modal displacements are insignificant compared to quasi-static values while for Humber, lateral vibrations in wind are significant, as the study of extreme wind events will later show. These values do not reflect the overall ranges due to quasi-static components; however,

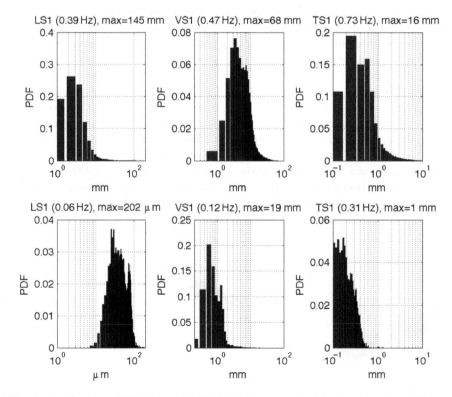

Figure 11. Probability density function of 30-min modal RMS displacements for Humber (upper) and Tamar (lower) bridges for fundamental lateral (L), vertical (V) and torsional (T) symmetric (S) modes. Note the different units for Tamar LS1.

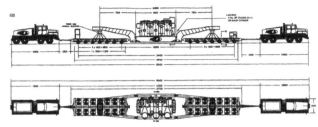

Figure 12.   Passage of a 300 tonne vehicle (65 m), westbound on Tamar Bridge, 31 October 2010.

for random vibration of approximately constant level during the 30-min averaging period, the peak factor scaling RMS to maxima is approximately three.

6.   Extreme response events

The most revealing performance information derives from extreme (operational) loading events, e.g. strong winds, low or high temperatures, weak or strong sunlight and exceptional vehicle loads. Significant temperature effects on Tamar Bridge deformation have already been shown via the correlation plots and the only noticeable effect of wind has been increased acceleration response in the lower vibration modes for mean wind speeds exceeding 10 m/s. Hence, the main interest with wind is at Humber Bridge where aero-elastic wind–structure interaction results in observable changes in modal properties (Diana, Cheli, Zasso, Collina, & Brownjohn, 1992) as well as significant dynamic and quasi-static deformations.

Both structures have experienced extreme traffic loads, usually due to multiple vehicles in traffic jams. Tamar Bridge experienced a single 300 tonne vehicle in

2010 (Westgate, Koo, Brownjohn, & List, 2014), while Humber regularly experiences abnormal loads of around 100 tonnes.

6.1   Heavy vehicle crossing

The single most significant loading event during the monitoring of Tamar Bridge was the passage of a two-tractor low-loader carrying a power station component, with total vehicle weight of 300 tonnes (Figure 12). To track the vehicle, the RTS was reconfigured to track a single target at the fastest rate possible, while a second standalone RTS (Leica TS30) was deployed to track Saltash Tower top deflections, also using a single reflector. Other response parameters such as extension, cable tension and acceleration were simultaneously recorded.

Experimentation with the two RTS units determined that they could operate at sample rates up to 3 Hz. Saltash expansion joint extension and stay cable tension data were interpolated at this rate and merged with the RTS data to produce the plots of Figure 13. The whole vehicle crossing event lasted 3.5 min and the recorded peak mid-span

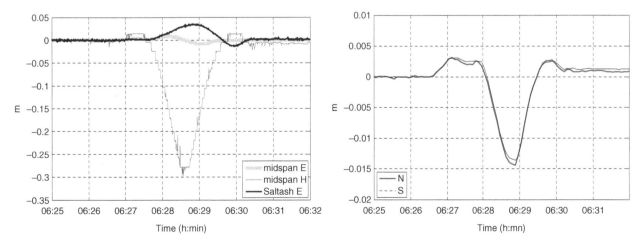

Figure 13.   Passage of a 300 tonne vehicle, 31 October 2010: RTS (left) and extensometer (right) data. Extension increases as deck moves away (to east) from Saltash Tower.

vertical deflection of 300 mm corroborated the authors' and the consultant's predicted values.

For Humber Bridge, the heaviest vehicle was observed during the 1990 monitoring campaign (Brownjohn et al., 1994). The 172 tonne vehicle generated a mid-span vertical deflection of just over 400 mm, in fact smaller than deflections due to strong winds and normal temperature variations. From the present campaign, Figure 14 shows the clear effects of heavy vehicles crossing in both directions via the main span deformations captured using several sensor types and recorded during low-wind conditions. The largest deformations during this interval are due to a 110 tonne vehicle, logged by an independent off-bridge weigh-in-motion station.

The vertical deformation from mid-span GPS is clear, as are the rotations provided by differencing the GPS heights across the deck. The rotations are corroborated by horizontal accelerometer and Nivel inclinometer signals converted to height differences across the same separation as the GPS antenna. A notable effect, similar to that observed at Tamar Bridge is the corresponding northerly movement of the main span followed by a southerly over-correction for southbound vehicles (and vice-versa). This effect has contributed strongly to wear of Humber A-frame bearings.

6.2 Traffic jam

Traffic jams also produce large and anomalous deformations and can be of particular concern to bridge operators when they coincide with extreme temperatures. Figure 15 shows deck rotations derived from DC component of lateral acceleration and changes in stay cable tension during one event on Tamar Bridge in April 2011 when neither the level system nor the RTS were in operation. Tension variations in the two stay cable pairs normally go in opposite directions due to deck expansion (as in the middle plot as the day progresses) but in this case their simultaneous increase around 9 am appears as a clear anomaly. The automated modal analysis system also detected noticeable dips in mode frequencies due to the added mass of the vehicles.

For Humber Bridge, traffic jams are uncommon due to low traffic volume, but one event on 6 April 2012 (Figure 16) was clearly identified by anomalous GPS height signals, appearing as 0.5 m shifts in 30-min average values for both sensors. The 1 Hz-sampled raw time series show mid-span vertical deflection as well as a rotation (west side dipped) for about 25 min caused by an accident blocking the northbound side of the bridge – while the southbound lanes were open (with spikes caused by heavy vehicles).

6.3 Extreme winds at Humber

The greatest static and dynamic deformations recorded at Humber Bridge have been due to wind. Examples are shown for two strong wind events, on 23 May 2011 (maximum mean wind speed 20 m/s, maximum gust 30 m/

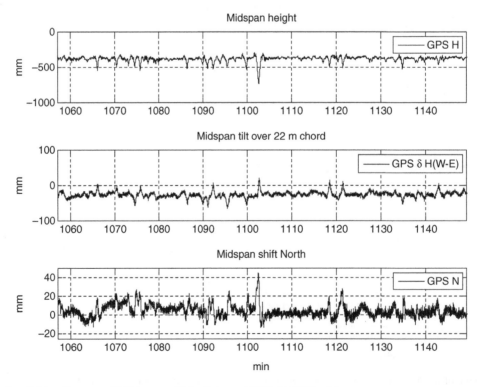

Figure 14.   Humber Bridge. Passage of 110 tonne vehicle, 7 August 2012 (at 1102 min).

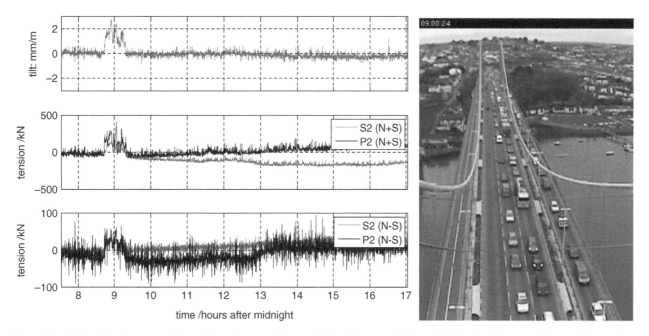

Figure 15. Tilt and main span cable tension changes at Tamar Bridge due to traffic jam (right) of vehicles travelling to Plymouth.

s) and 30 January 2013 (the 26 m/s mean speed and 39 m/s gust), both for westerly winds (which dominate in the UK).

The event of 23 May (Figure 17) is characterised by a rising wind speed producing a large mid-span lateral deflection followed by strong lateral oscillations. These are accompanied by vertical (downwards) deflections, rotations and longitudinal movement, with GPS northings inverted as 'southings' for comparison with extensions. The inclinometer is not effective at measuring rotations in strong winds since the low-range high-resolution option installed clips for angles greater 0.0025 rad (55 mm height difference). The event that happened on 30 January is exceptional, causing lateral oscillation up to 1 m amplitude and depression of the box deck by 1 m.

All signals were available for the May 2011 event, but the east GPS signal was not available for the January 2013

event, while accelerometer data do not cover the complete event. For both events, the extensometer signals are particularly interesting: the difference between the two sensors on east and west sides of the Hessle Tower tracks the lateral deflection perfectly, but is approximately one-tenth of the lateral deflection. In the figure, these differences are superimposed on changes in the averaged extension readings that for 30 January appear as a reduction of 0.2 m and which is reflected by a 0.2 m increase in GPS northings. In other words, the whole span shifted north 0.2 m, an effect confirming the observation in Figure 9.

Total accelerations in any direction do not translate simply to displacements but during the two strong wind events, mid-span acceleration response reached $0.2\,\mathrm{m/s^2}$ vertical amplitude and $0.05\,\mathrm{m/s^2}$ lateral amplitude.

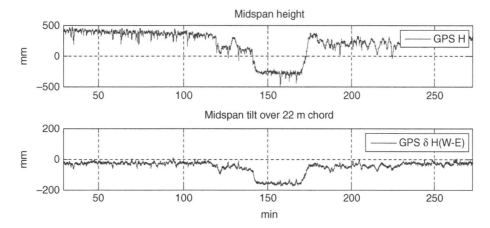

Figure 16. Vertical deformation and rotation at Humber Bridge due to traffic jam of northbound vehicles.

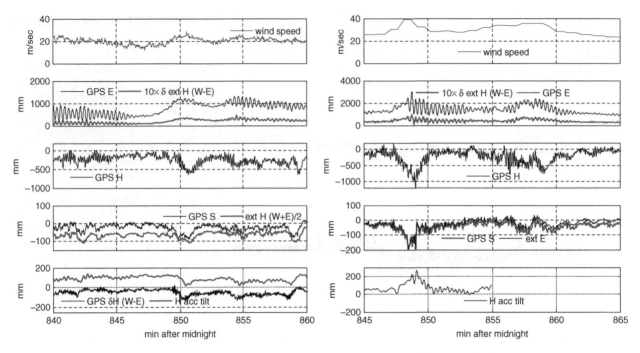

Figure 17.   Humber Bridge. Time series of strong-wind response for 23 May 2011 and 30 January 2013.

7.   Humber Bridge main span bearing movements

Main and side spans of Humber Bridge are constrained against lateral and vertical translation and against rotation about the longitudinal axis by means of a pair of A-frame rockers (Figure 18) at each end, while roadway continuity is via Demag expansion joints. The combination of frequencies is designed to allow free axial movement due to vehicles, wind and temperature.

7.1   Bearing influence on dynamic performance

The operation of the bearings is known to have a strong influence on the dynamic properties of Humber Bridge. The validated finite element model (Brownjohn et al., 1987) demonstrated that with all bearings behaving as perfect sliding and pinned joints, the first vertical vibration mode would be anti-symmetric, with frequency 0.108 Hz, whereas with either end not free to slide the first mode is the 'symmetric' one. In reality, the symmetric mode always appears first at 0.116 Hz, with the anti-symmetric mode at 0.15 Hz during all vibration measurements since 1985.

For both Humber Bridge and Tamar Bridge, automated operational modal analysis operates to provide modal parameters for the lowest few vertical, lateral and torsional modes. For both bridges, the first lateral (symmetric) modes follow diurnal patterns with ranges up to 20% of mean values, but the causes are not obvious since temperature, wind and traffic all have diurnal variations. For Tamar Bridge, the effect has been shown to be strongly

dependent on traffic load (Cross, Koo, Brownjohn, & Worden, 2013).

For Humber Bridge, the effect appears to be amplitude and wind-speed dependent. Modal frequencies and damping ratios for a range of first lateral mode (LS1) RMS response levels are shown in Figure 19. It is clear that modal frequency stabilises at 0.054 Hz for high amplitude response. Likewise, for strong response, damping values drop below 1% while at low response levels damping can exceed 10%. The frequency variation/ duality for mode LS1 was previously observed in 1988 by Building Reseach Establishment (Littler, 1992). It is believed that the damping and frequency trends are consistent with high friction at the bearings that inhibit the differential bearing movement that is linked to lateral deformation of the deck shown in Figure 17.

7.2   Bearing cumulative movement

The visible condition of the bearings as shown in Figure 18, and which by the argument above has not affected the global behaviour of the bridge, led to a requirement to replace them in 2013 (Hornby, Collins, Hill, & Cooper, 2012). Extension data for a period in 2011 were used to inform design through information on the exected travel of the bearing rockers and the bearing replacement is expected to have completed in 2014. Figure 20 indicates the extent of this motion for 1 day each of windy and calm conditions. The slow variations in the extension time series are due to temperature, the faster

Figure 18. Hessle Tower-main span west A-frame rocker (left) and close up of each of the two lower bearings showing different wear. The gap is has disappeared on one bearing due to excessive wear.

variations being due to quasi-static wind response then vehicle-induced motion (as in Figure 14), then low frequency vertical and lateral modes.

Clearly, strong winds can double the travel, but even for calm days, 50 m of accumulated daily travel amounts to 580 km over the bridge lifespan to date, probably a

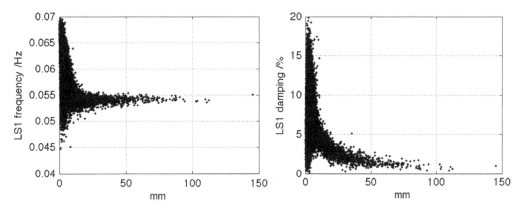

Figure 19. Frequency and damping estimates for fundamental lateral mode LS1 as a function of modal response. From 30-min averages.

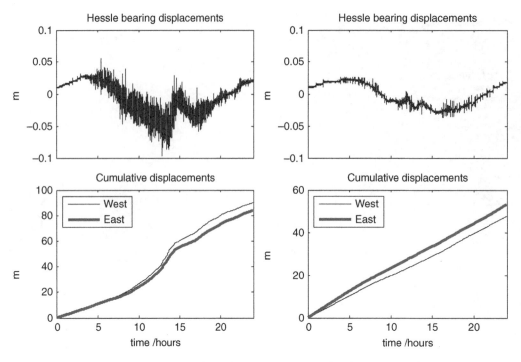

Figure 20. Humber Bridge. Bearing movements on windy (23 May 2011) and calm day (19 June 2011).

little more than anticipated and going towards explaining the bearing condition before replacement. Despite this, there appears to have been no significant change in the effect of bearings on modal properties; first vertical mode frequencies have not changed. Also the quasi-static component of bearing travel is very similar between 1990/1991 monitoring (some data from that study are still available) and the recent measurements. Apparently, the bearings have been forced to endure the bridge motion, regardless of consequence and the behaviour seems likely not to have changed significantly over the bridge lifetime.

7.3 Response in frequency domain

Time series of response for a day of moderate winds have been converted to frequency domain to illustrate the performance of different sensors in capturing the quasi-static and dynamic response of Humber Bridge. The result is Figure 21 which covers the frequency range 0.03–2 Hz with accelerometer signals converted to displacements (through amplitude division by squared circular frequency). Going from left to right, the plots illustrate vertical, lateral and rotational deformations. GPS data, which are effectively noise above 0.5 Hz due to interpolation from the 1 Hz original sample rate, are really only useful for capturing the first vertical, lateral and torsional modes (VS1, LS1, TS1), and accelerometers do a much better job for dynamic response with clear identification of higher modes.

It is educational that accelerometers capture vertical motion successfully down to 0.03 Hz and in fact even down to 0.01 Hz. Lateral accelerometer, lateral GPS and inclinometer all pick up the first lateral mode (0.054 Hz). The inclinometer ('inc') also picks up some higher lateral modes (e.g. 0.4 Hz) seen in the accelerometer signal, showing that it is not decoupled from inertial effects. On the other hand, it does not capture the first torsional mode TS1 (0.310 Hz) well. GPS height difference WH − EH (δH (W − E) in other figures) hardly captures any dynamic response.

7.4 GPS vs. accelerometers vs. inclinometers

Figures 14, 17 and 21 show the potential for estimating deformations using cheaper or more convenient sensors. For example, the inclinometer provides practically the same data as the DC component of the lateral accelerometer but little else. Similarly knowing (as a result of this exercise) that height differences between GPS 'rover' receivers either side of the bridge could be estimated using a single rover and the DC component of the lateral acceleration, the second rover appears to be redundant.

There is a strong correlation between lateral movement of the deck and rotation (about a vertical axis) at the deck bearings, but the relationship depends on the deformed shape along the deck at high amplitudes when it is presumed that the bearings are 'unstuck'. In any case, the GPS sensors have so far been more reliable than extensometers. Hence, a combination of single GPS, DC accelerometer and a pair of extensometers is about the minimum requirement for capturing quasi-static deck response. A pair of vertical accelerometers completes the picture with dynamic response.

Clearly, accelerometers can capture dynamic components of response down to frequencies below the lateral mode frequency (0.054 Hz), so a relatively expensive GPS rover is an extravagant means for recovering dynamic response. Likewise, use of high frequency GPS data (e.g. above 2 Hz) to recover dynamic response appears to be pointless, although high frequency capability is improved using averaging and high frequency receivers. For example, displacement amplitude of 5 mm, which is a reasonable estimate of the best resolution capability of GPS, translates to an acceleration of 0.8 m/s^2 at 2 Hz, well above acceleration levels observed in the low frequency modes for both bridges.

The low frequency response of the accelerometer illustrated in Figure 21 was tested to see how well it could reproduce quasi-static vertical motion. Figure 22 shows an attempt using 23 May 2011 data and a high pass filter set at 0.0025 Hz for the double integration of averaged east and west signals. The right plot compares dynamic components of the signals by applying 0.08 Hz high pass filters. The timing mismatch could be in part due to use of

slightly different filtering and in part due to data acquisition PC clock drift, but the match is convincing and the small differences obvious.

8. Overview of performance

Comparison of two bridges and a range of sensors permits the authors to comment on the similarities and differences and the implications for bridge performance monitoring.

8.1 Sensor performance

The similarities and differences between the two systems, and their contemporaneous operation provide insights into best approaches for monitoring such structures considering resolution, noise and frequency content as well as cost and operation. In the higher frequency range, certainly up to no more than 5 Hz and usually not exceeding 1 Hz for a long-span bridge, information about dynamic (modal) characteristics is relevant to academics but only to a minority of high-end specialist consultants and very well-informed bridge operators. Dynamic characterisation serves a purpose in the role of calibrating simulations of dynamic behaviour in unusual events such as earthquakes and wind storms and in very limited circumstances can help to identify altered structural condition. At present, however,

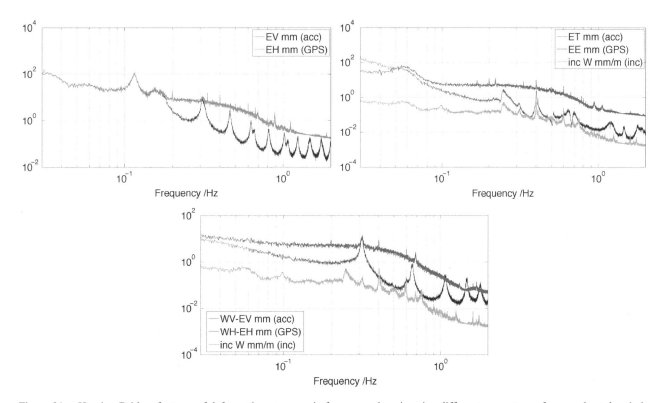

Figure 21. Humber Bridge, features of deformation response in frequency domain using different sensor types for a moderately windy day, 7 October 2011. Left: Vertical displacement by accelerometer and GPS. Middle: Lateral displacement from accelerometer and GPS plus rotation angle from inclinometer. Right: Rotational displacement from accelerometer and GPS differences plus rotation angle from inclinometer.

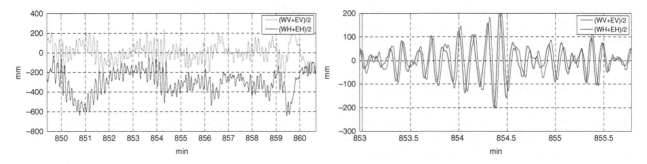

Figure 22. Humber Bridge. Comparison of integrated acceleration (V) and GPS (H) for 23 May 2011. The left side has double-integrated acceleration with low-pass filter at 0.0025 Hz and right side with low-pass filter at 0.08 Hz (leaving only dynamic components).

deformations in the sub-modal, quasi-static frequency range tend to be of greatest direct interest as they relate to serviceability in normal operation and are the parameters that designers aim to limit. For some types of response, deformations are small and sensor resolution (or threshold) is important, as are noise characteristics.

Noise performance of accelerometers is relatively straightforward to describe. The QA750 accelerometer is reported as having 'resolution/threshold' better than 1 μg, while our own studies (Brownjohn & Botfield, 2009) have shown that broadband noise floor (0–100 Hz) is 3 μm/s^2/$\sqrt{Hz}$ or better. In terms of displacement, these figures convert to 2.5 mm resolution and 0.75 mm/$\sqrt{Hz}$ noise at 0.01 Hz and to 0.25 μm and 0.075 μm/$\sqrt{Hz}$ noise at 1 Hz. For the Nivel220 inclinometer, in the range setting used here, the resolution is 1 μrad and accuracy 47 μrad (no noise value is quoted by Leica). For the QA750 horizontal accelerometers used in both bridges to indicate tilt, output due to main span rotation θ is resolved gravity (g) component $g \sin\theta$, so resolution is 1 μrad and noise 0.3 μrad/Hz.

For ASM W12 mechanical extensometers used at Tamar, electrical noise is quoted which translates to $\sigma = 0.16$ mm on extensions, while the Hilti PD4 units at Humber have quoted accuracy of ± 2 mm. So while the lasers are more convenient, they are less accurate. More accurate lasers are available but they cannot operate at the 7 m distance used at Humber Bridge. Accuracy of the Leica TCA1201M is also clear: 2 mm + 2 ppm for the electronic distance measurement and 1 arcsec for the theodolite. Hence, at the 650 m distance to the furthest reflector, accuracy is 3.3 mm.

For the GPS, noise, resolution and accuracy are harder to characterise. Nickitopoulou, Protopsalti, and Stiros (2006) suggest standard accuracy of 15 mm for horizontal measurements and 35 mm for vertical measurements, both at 98.5 percentile level, without gross errors such as cycle slip or multipath. Assuming Gaussian distribution (and GPS noise has unusual characteristics), the standard deviation is still more than 5 mm horizontally and 10 mm vertically. The previous short-term Humber Bridge GPS

campaign (Ashkenazi & Roberts, 1997) provided a figure of 1 mm resolution in horizontal plane and 3 mm in the vertical direction. The differences in Figure 22 suggest a poorer resolution, yet the GPS signals still reflect the scale of the bridge motion, and the quality of the data is good enough to provide clear correlations with other load or response parameters.

One way to check GPS performance at Humber would be to examine signals which should be practically identical, e.g. the pairs of northing and easting signals from the east and west receivers. The difference of eastings (sway) for the same data-set used for Figure 21 shows typical modal response at 0.22, 0.37 and 0.49 Hz. None of these frequencies correspond to lateral modes, so it is possible that they correspond to 'cable modes' where only the suspension cables participate, the lowest of which was predicted at 0.24 Hz in the first three-dimensional analytical model study of the bridge (Dumanoglu & Severn, 1987). Even with the weak modal response, the two signal differences have $\sigma = 2$ mm in the 0.2–0.5 Hz band.

The 0.15 m range of vertical deformations at Tamar Bridge seems to have validated original advice against using GPS but a direct comparison would still be useful. For Humber Bridge, the range of vertical movements has been 2 m, slightly larger for the lateral direction, and all values are for 30-min averages. The major advantage of GPS is the uniform sampling rate set at 1 Hz to capture quasi-static effects of vehicle and wind and provide a cross-check on accelerometer data for lower frequency vibration modes. For both bridges, double integration could not capture the quasi-static vehicle-induced vertical deformations, and when tracking multiple targets at Tamar, the RTS sample rate is far too slow.

The quasi-static tracking solution for road bridges spans shorter than Tamar Bridge would be either a single target RTS (as deployed for tracking the heavy vehicle) or a purely optical tracking system, which would depend on weather conditions. For longer bridges, GPS is ideal since RTS and other optics-based systems suffer with poor visibility conditions and also degrade with distance.

Accelerometers would measure rotation and capture modal response while GPS/RTS would overlap with lower modes and resolve quasi-static response. Hence, accelerometer and GPS signals linked, e.g. by Kalman filter should be a good approach (Smyth & Wu, 2007). Neither extensometer system used here was totally satisfactory; thus a solution with short-range, high-resolution lasers would be preferred, preferably at both ends of the span, at either side.

Strain gauges would not have been useful for these deformation measurements since multiple strain measurements (or assumed uniform strain) would be necessary to estimate, e.g. Humber box deck expansion or curvature, and such data were available directly from the deformations. The only application in this study was in the 'load cells' for the Tamar additional stay cables, which carry about 10% of the weight of the suspended structure.

8.2   Bridge performance

Tamar Bridge seems relatively immune to extreme weather due to short truss span (reduced wind effect) and more benign temperature climate. Road traffic is much heavier than at Humber and there is also interest in bearing performance due to the effects of the strengthening and widening operation. For this bridge, the normal operation is probably the greater interest. Humber Bridge is the more vulnerable to wind effects with some very large deformations, and suffers more extreme temperatures, albeit so far with no adverse effects. Regarding safety, this study provides no surprises in the way of unexpected or worrying performance, and the monitoring systems are most useful for providing information to support retrofit that for these bridges relates to bearings and deck surfacing.

9.   Conclusions

Concerning bridge deformations, for both bridges the major driver under normal circumstances is temperature, which operates over the slowest timescale of all forms of loading. The deformation patterns induced by variation of temperature in time and position require a combination of sensors to identify them, with focus on movements in the vertical plane. The largest deflections have been due to rare extreme wind events and abnormal vehicles, while deflections due to normal heavy vehicle and strong wind loads at sub-dynamic quasi-static frequencies are smaller than the temperature-induced motions. In all but exceptional winds, deformations in individual vibration modes are relatively small.

Identifying these deformations can be demanding for permanent instrumentation systems. GPS appears to be a standard choice for long-span bridges, but below a certain span length it is degraded compared to optics-based systems with 3D capability, of which the present best solution is the total station. Accelerometers provide a wealth of data limited only by ability to capture quasi-static response and so investment in high-quality devices is worthwhile, while laser or LED-based devices for line of sight measurements should be best suited for the important role of tracking bearing motion.

Notes

1.   Email: k.y.koo@exeter.ac.uk
2.   Email: andrew@ishm.co.uk
3.   Email: david.list@tamarcrossings.org.uk

References

Aktan, A.E. (2011). *International bridge study workshop.* Retrieved from: http://www.cait.rutgers.edu/system/files/u5/Aktan_IBS_slides_061311.pptx

Ashkenazi, V., & Roberts, G.W. (1997). Experimental monitoring of the Humber Bridge using GPS. *Proceedings Institution of Civil Engineers, Civil Engineering, 120,* 177–182.

Brownjohn, J.M.W., Bocciolone, M., Curami, A., Falco, M., & Zasso, A. (1994). Humber Bridge full-scale measurements campaigns 1990–1991. *Journal of Wind Engineering and Industrial Aerodynamics, 52,* 185–218.

Brownjohn, J.M.W., & Botfield, T. (2009). A folded pendulum isolator for evaluating accelerometer performance. *Experimental Techniques, 33,* 33–37.

Brownjohn, J.M.W., Dumanoglu, A.A., Severn, R.T., & Taylor, C.A. (1987). Ambient vibration measurements of the Humber Suspension Bridge and comparison with calculated characteristics. *Proceedings Institution of Civil Engineers Part 2, 83,* 561–600.

Brownjohn, J.M.W., Magalhães, F., Caetano, E., & Cunha, A. (2010). Ambient vibration re-testing and operational modal analysis of the Humber Bridge. *Engineering Structures, 32,* 2003–2018.

Calçada, R., Cunha, A., & Delgado, R. (2005). Analysis of traffic-induced vibrations in a cable-stayed bridge. Part 1: experimental assessment. *Journal of Bridge Engineering, 10,* 370–385.

Catbas, F.N., Correa-Kijewski, T., & Aktan, A.E. (Eds.). (2013). *Structural identification of constructed systems. Approaches, methods and technologies for effective practice of St-Id.* Reston, VA: American Society of Civil Engineers.

Cross, E.J., Koo, K.-Y., Brownjohn, J.M.W., & Worden, K. (2013). Long-term monitoring and data analysis of the Tamar Bridge. *Mechanical Systems and Signal Processing, 35,* 16–34.

Diana, G., Cheli, F., Zasso, A., Collina, A., & Brownjohn, J.M.W. (1992). Suspension bridge parameter identification in full-scale test. *Journal of Wind Engineering and Industrial Aerodynamics, 41,* 165–176.

Diana, G., Falco, M., Cheli, F., & Cigada, A. (2003). The aeroelastic study of the Messina Straits Bridge. *Natural Hazards, 30,* 79–106.

Dumanoglu, A.A., Brownjohn, J.M.W., & Severn, R.T. (1992). Seismic analysis of the Fatih Sultan Mehmet (Second Bosporus) Suspension Bridge. *Earthquake Engineering and Structural Dynamics, 21,* 881–906.

Dumanoglu, A.A., & Severn, R.T. (1987). Seismic response of modern suspension bridges to asynchronous vertical ground motion. *Proceedings Institution of Civil Engineers Part 2*, *83*, 701–730.

Erdoğan, H., & Gülal, E. (2011). Ambient vibration measurements of the Bosphorus Suspension Bridge by total station and GPS. *Experimental Techniques*, *37*, 16–23.

Hjorth-Hansen, E., & Niggard (1977). *Et akselerometer og dets signalkjede* [An accelerometer and its signal chain]. Trondheim: NTH.

Hornby, S.R., Collins, J.H., Hill, P.G., & Cooper, J.R. (2012). Humber Bridge A-frame refurbishment/replacement. In *Proceedings of the 6th international conference on bridge maintenance, safety and management (IABMAS)*, Stresa, Italy (pp. 3170–3177). London: Taylor & Francis Group.

Kim, S., Pakzad, S.N., Culler, D., Demmel, J., Fenves, G., Glaser, S., & Turon, M. (2007). Health monitoring of civil infrastructures using wireless sensor networks. In *Proceedings of the Sixth international symposium on information processing in sensor networks (IPSN2007)* (pp. 254–263). New York: Association for Computing Machinery.

Kite, S., Carter, M., & Hussain, N. (2010). Forth replacement crossing – Design for safe maintenance and management. In *Proceedings of the IABMAS2010, fifth international conference on bridge maintenance, safety and management*. Philadelphia, PA (pp. 1826–1833). London: Taylor & Francis Group.

Ko, J.M., & Ni, Y.Q. (2005). Technology developments in structural health monitoring of large-scale bridges. *Engineering Structures*, *27*, 1715–1725.

Koh, H.-M., Lee, H.-S., Kim, S., Choo, J.F., Boller, C., Chang, F.K., & Fujino, Y. (2009). Monitoring of bridges in Korea. In *Encyclopedia of structural health monitoring* (pp. 1–23). Chichester: Wiley.

Littler, J.D. (1992). Ambient vibration tests on long span suspension bridges. *Journal of Wind Engineering and Industrial Aerodynamics*, *41–44*, 1359–1370.

Macdonald, J.H.G., Dagless, E.L., Thomas, B.T., & Taylor, C.A. (1997). Dynamic measurements of the Second Severn Crossing. *ICE Proceedings Bridge Transport*, *123*, 241–248.

Marecos, J. (1978). The measurement of vertical displacements through water levelling method. *Materials and Structures*, *11*, 361–370.

Marecos, J., Castanheta, M., & Trigo, J.T. (1969). Field observation of Tagus River suspension bridge. *Journal of Structural Engineering*, *95*, 555–583.

Myrvoll, F., Di Biagio, E., & Hansvold, C. (1994). Instrumentation for monitoring the Skarnsundet cable-stayed bridge. In *Third Symposium on straits crossings* (pp. 207–215). Alesund: Balkema.

Nickitopoulou, A., Protopsalti, K., & Stiros, S. (2006). Monitoring dynamic and quasi-static deformations of large flexible engineering structures with GPS: Accuracy, limitations and promises. *Engineering Structures*, *28*, 1471–1482.

Pakzad, S.N., Fenves, G.L., Kim, S., & Culler, D.E. (2008). Design and implementation of scalable wireless sensor network for structural monitoring. *Journal of Infrastructure Systems*, *14*, 89–101.

Rainer, J.H. (1985). *Effect of rotation on motion measurements of towers and chimneys*. Ottawa: DBR NRC Building Research Note, no. 230.

Ren, W.-X., Harik, I.E., Blandford, G.E., Lenett, M., & Baseheart, T.M. (2004). Roebling suspension bridge. II: Ambient testing and live-load response. *Journal of Bridge Engineering*, *9*, 119–126.

Roberts, G.W., Meng, X., Brown, C.J., & Andrew, A. (2006). Measuring the movements of the Forth Road Bridge by GPS; lorry trials. In A. Kumar, C.J. Brown, & L.C. Wrobel (Eds.), *The first international conference on advances in bridge engineering. Bridges – Past, present and future*, Uxbridge (pp. 28–36). Uxbridge: Brunel University.

Sloan, T.D., Kirkpatrick, J., Boyd, J.W., & Thomson, A. (1992). Monitoring the in service behaviour of the Foyle Bridge. *The Structural Engineer*, *70*, 130–134.

Smyth, A., & Wu, M. (2007). Multi-rate Kalman filtering for the data fusion of displacement and acceleration response measurements in dynamic system monitoring. *Mechanical Systems and Signal Processing*, *21*, 706–723.

Stephen, G.A., Brownjohn, J.M.W., & Taylor, C.A. (1993). Measurements of static and dynamic displacement from visual monitoring of the Humber Bridge. *Engineering Structures*, *15*, 197–208.

Stiros, S., & Psimoulis, P.A. (2012). Response of a historical short-span railway bridge to passing trains: 3D deflections and dominant frequencies derived from robotic total station (RTS) measurements. *Engineering Structures*, *45*, 362–371.

Sumitro, S. (2001). Current and future trends in long span bridge health monitoring system in Japan. In *Proceedings of the seventh international symposium on smart structures and materials*, SPIE, Newport, CA, March 4–8, 2001. Bellingham, WA: SPIE.

University of Washington (1954). *Aerodynamic stability of suspension bridges with special reference to the Tacoma Narrows bridge*. Bulletin No. 116 University of Washington Engineering Experiment Station. Seattle: University of Washington.

Westgate, R., Koo, K.-Y., Brownjohn, J.M.W., & List, D. (2014). Suspension bridge response due to extreme vehicle loads. *Structure and Infrastructure Engineering*, *10*, 821–833.

Wong, K.-Y. (2007). Design of a structural health monitoring system for long-span bridges. *Structure and Infrastructure Engineering*, *3*, 169–185.

Zasso, A., Vergani, M., Bocciolone, M., & Evans, R. (1993). Use of a newly designed optometric instrument for long-term, long distance monitoring of structures, with an example of its application on the Humber Bridge. In *Proceedings of the 2nd international conference on bridge management*. Guildford (pp. 1–10). London: Telford.

Zhou, H.F., Ni, Y.Q., & Ko, J.M. (2006). Analysis of structural health monitoring data from the suspension Jiangyin bridge. In A. Guemes (Ed.), *Proceeding of the 3rd European workshop on structural health monitoring* (pp. 364–371). Lancaster, PA: DEStech.

Structural health monitoring of a cable-stayed bridge with Bayesian neural networks

Stefania Arangio and Franco Bontempi

In recent years, there has been an increasing interest in permanent observation of the dynamic behaviour of bridges for long-term monitoring purpose. This is due not only to the ageing of a lot of structures, but also for dealing with the increasing complexity of new bridges. The long-term monitoring of bridges produces a huge quantity of data that need to be effectively processed. For this purpose, there has been a growing interest on the application of soft computing methods. In particular, this work deals with the applicability of Bayesian neural networks for the identification of damage of a cable-stayed bridge. The selected structure is a real bridge proposed as benchmark problem by the Asian-Pacific Network of Centers for Research in Smart Structure Technology (ANCRiSST). They shared data coming from the long-term monitoring of the bridge with the structural health monitoring community in order to assess the current progress on damage detection and identification methods with a full-scale example. The data set includes vibration data before and after the bridge was damaged, so they are useful for testing new approaches for damage detection. In the first part of the paper, the Bayesian neural network model is discussed; then in the second part, a Bayesian neural network procedure for damage detection has been tested. The proposed method is able to detect anomalies on the behaviour of the structure, which can be related to the presence of damage. In order to obtain a confirmation of the obtained results, in the last part of the paper, they are compared with those obtained by using a traditional approach for vibration-based structural identification.

Introduction

In recent years, there has been an increasing interest in long-term monitoring of bridges. This depends on various factors. First of all, the research community has been alerted by some tragic events and collapses of bridges that pointed out the vulnerability of some existing structures and the uncertainty in their analysis for monitoring and maintenance purpose (Biondini, Frangopol, & Malerba, 2008; Bontempi & Giuliani, 2010; Crosti, Duthinh, & Simiu, 2011; Olmati & Gentili, 2010). In this sense, data collected on site are essential both for checking the accomplishment of the expected performance during their entire life-cycle and for validating the original design (Arangio, 2012; Koh, Kim, Lim, Kang, & Choo, 2010; Petrini & Bontempi, 2011).

In addition, recent technological advances have contributed to make the installation and operation of permanent dynamic monitoring systems more practical and economical, and nowadays, it is possible to have real-time data collecting and remote control systems (Arangio et al., 2014; Arangio & Bontempi, 2014). In order to profit from the latest technological developments and to have a system that is truly effective for the assessment of the structural condition, a continuous online processing of the collected data is required.

Therefore, a great effort in research has been devoted to the development of effective methods for processing large quantity of monitoring data (see, for example, Ko, Ni, Zhou, Wang, & Zhou, 2009; Li et al., 2006) and developing efficient strategies for maintenance (Bontempi, Gkoumas, & Arangio, 2008; Casas, 2010; Frangopol, Saydam, & Kim, 2012). Various methods based on soft computing techniques, as neural networks and fuzzy logic, have proved to be very efficient (see, for example, Adeli, 2001; Arangio & Bontempi, 2010; Ceravolo, De Stefano, & Sabia, 1995; Choo & Koh, 2009; Dordoni, Malerba, Sgambi, & Manenti, 2010; Freitag, Graf, & Kaliske, 2011; Kim, Yoon, & Kim, 2000; Ko et al., 2002; Sgambi, Gkoumas, & Bontempi, 2012; Tsompanakis, Lagaros, & Stavroulakis, 2008) and have attracted the attention of the researchers community. In particular, this work deals with the applicability of the Bayesian neural networks for monitoring the structural integrity of a cable-stayed bridge.

The selected structure is a real bridge, the Tianjin Yonghe Bridge, proposed as benchmark problem by the Asian-Pacific Network of Centers for Research in Smart Structure Technology (ANCRiSST SHM Benchmark Problem, 2011) (Figure 1). They shared data coming from the long-term monitoring of the bridge with the structural

Figure 1. Tianjin Yonghe Bridge considered for the ANCRiSST benchmark problem (picture by Liuzhou OVM Machynery Co., Ltd, http://www.ovmchina.com/ovmweb-e/Default.aspx, accessed January 2013).

health monitoring community in order to assess the current progress on damage detection and identification methods on a full-scale example. The data set includes vibration data (time series of the accelerations of the deck) before and after the bridge was damaged; this is a rare case of an instrumented bridge that has been damaged. The available time series have been used to test a Bayesian neural networks-based strategy for damage detection. The proposed strategy is an extension of a method proposed by Arangio and Beck (2012) that evaluated the possibility to identify the occurrence and the location of damage by analysing the time series of the structural responses under ambient vibrations. They worked with data obtained from numerical models. On the other hand, the present application is based on the processing of real noisy continuous ambient measurements.

In the first part of the paper, the bridge of the ANCRiSST benchmark problem is described and the Bayesian neural networks are briefly presented; in the second part, the procedure for damage detection is applied to the bridge; in the last part, the results obtained with the proposed approach are cross validated with those obtained by using a traditional approach for vibration-based structural identification in the frequency domain.

The ANCRiSST benchmark problem

The Asian-Pacific Network of Centers for Research in Smart Structure Technology (ANCRiSST – http://smc.hit.

edu.cn/ancrisst2011/) is a consortium that has been created with the purpose of assessing the current progress of smart materials and structures technology and developing synergies among researchers in various disciplines from different countries that will facilitate joint research projects that cannot be easily carried out by the individual centres. This consortium was established in 2002 and currently consists of 20 research institutions. From October 2011, they opened for researchers in the structural health monitoring community a benchmark problem based on a real bridge (ANCRiSST SHM Benchmark Problem, 2011). The available data set consists of vibration data that have been obtained before and after the bridge was damaged, thus can serve as a full-scale benchmark to evaluate and further develop structural damage detection and identification methods.

Description of the bridge

The Tianjin Yonghe Bridge is one of the earliest cable-stayed bridges constructed in Mainland China. It has two towers, a main span of 260 m and two side spans of 25.15 + 99.85 m each. The full width of the prestressed concrete deck is about 13.6 m, including a 9-m roadway and sidewalks. The elevation with the main dimensions is in Figure 2. In Figure 3, it is shown a picture of the ANSYS finite element model (FEM) of the bridge, which was shared by the consortium and which was at the base of the

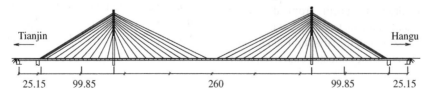

Figure 2. Skyline of the bridge with the main dimensions.

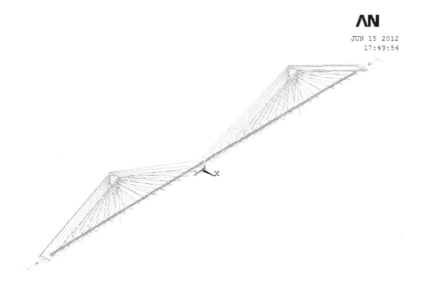

Figure 3. FEM model.

numerical analysis carried out in this work. More details on the bridge are given in Li, Li, Ou, and Li (2010).

The Tianjin Yonghe Bridge was opened to traffic since December 1987. After 19 year of operation, cracks as wide as 2 cm were found in the bottom of the segment over the mid-span and the cables, especially at the anchors, were seriously corroded. For this reason, significant mainten-ance works were carried out: the deck girder over the mid-span was retrofitted and all the stay cables were replaced. For ensuring the future safety of the bridge, a sophisticated SHM system has been designed and implemented by the Research Center of Structural Health Monitoring and Control of the Harbin Institute of Technology and the time histories of the acceleration of the bridge deck and the towers started to be collected. More details on the monitoring system are given in Lan, Zhou, Sun, and Ou (2008). In August 2008, two different kinds of damage were detected during the bridge inspection. The external portions of both spans of the bridge were seriously cracked (damage scenario 1 – Figure 4). Meanwhile, the piers were damaged by overloading and the bridge experienced the partial loss of the vertical supports (damage scenario 2

– Figure 5). The same damages occurred simultaneous at the symmetric positions of the bridge.

Structural health monitoring system and available data set

The continuous monitoring system designed for the bridge includes 14 uniaxial accelerometers permanently installed on the bridge deck and 1 biaxial accelerometer that was fixed on the top of one tower to monitor its horizontal oscillation. An anemometer was attached on the top of the tower to measure the wind speed in three directions, and a temperature sensor was installed at the mid-span of the girder to measure ambient temperature (Figure 6). In particular, 7 of the 14 accelerometers were placed downstream and 7 upstream as indicated in Figures 6 and 7.

The data that were made public for the researchers regard both health and damaged conditions. Data in the health condition include time histories of the accelerations recorded by the 14 deck sensors and environmental information (wind and temperature). They consist in 24 data sets of 1 h recorded on 17 January 2008. The sampling frequency is 100 Hz. An example is shown in Figure 8. The second part of available data includes other measurements recorded at the same locations after some months, on 31 July 2008. In the meantime, as already said, some damages have been observed. The data set includes again time series of 1 h of the accelerations repeated for the 24 h at the same sampling frequency (100 Hz). After the detection of damage, four field tests have been carried out on the structure between 7 and 10 August 2008. The number of accelerometers has been increased to 18 for side and specific tests on both sides have been developed. In the present application, only data coming from the continuous monitoring have been used. The idea of the ongoing research is to use the data coming from the continuous

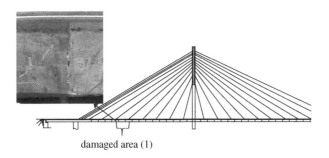

damaged area (1)

Figure 4. Damage scenario 1.

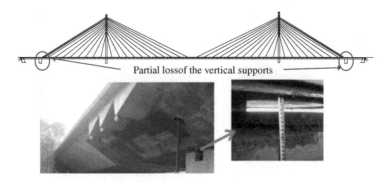

Figure 5. Damage scenario 2 (image from http://smc.hit.edu.cn).

monitoring system for the phase of damage detection. Once the damage is detected, in further studies, the more detailed measurements could be used for identifying the location of damage.

Bayesian neural networks for data processing

The considered time series included a large quantity of data, and it has been chosen to apply the neural networks for their processing. It was chosen to use neural networks because it has been proven that, if the response time history related to n time steps is known, they are a powerful tool for approximating the behaviour of the structures in the subsequent instants. Several applications of NNs used for such problems of function approximation and prediction showed that the neural network model is well suited for solving such problems (Lapedes & Farber, 1987; Qui & Zhang, 2003) because it is robust and fault tolerant and it can effectively deal with qualitative, uncertain and incomplete information.

A detailed explanation of the NNs is beyond the scope of this work but for a review on the subject, it is possible to refer to Adeli (2001), who illustrated the applications of neural networks to Civil Engineering during a decade, and to Waszczyszyn and Ziemianski (1999), who collected in a book various papers on the use of neural networks for the analysis and design of structures. As concerns more specifically the problem of damage identification and structural health monitoring, Ni, Wong, and Ko (2002) presented a two-stage neural network-based damage

detection method, where damage location is identified in a first stage and damage severity is estimated in a second stage; Ko, Sun, and Ni (2002) used neural networks in a multi-stage identification scheme for detecting damage in a cable-stayed bridge in Hong Kong; Xu and Humar (2006) presented a two-step algorithm that uses a modal energy-based damage index to locate the damage and a neural network technique to determine its magnitude.

In the following, the principles of the neural network model are briefly introduced, and its framework in the Bayesian point of view is explained. For more details, see for example Bishop (1995). The neural network concept has its origins in attempts to find mathematical representations of information processing in biological systems, but a neural network can also be viewed as a way of constructing a powerful statistical model for nonlinear regression. It can be described by a series of functional transformations working in different correlated layers (Bishop, 1995), that in the case of two layers, takes the form:

$$y_k(\mathbf{x}, \mathbf{w}) = h\left(\sum_{j=1}^{M} w_{kj}^{(2)} g\left(\sum_{j=1}^{D} w_{ji}^{(1)} x_i + b_{j0}^{(1)}\right) + b_{k0}^{(2)}\right), \quad (1)$$

where y_k is the kth output variable in the output layer; $\mathbf{x}$ is the vector of the D input variables in the input layer; $\mathbf{w}$ is the matrix including the adaptive weight parameters $w_{ji}^{(1)}$ and $w_{kj}^{(2)}$ and the biases $b_{j0}^{(1)}$ and $b_{k0}^{(2)}$ that are set during the training phase (the superscript refers to the considered layer); M is the total number of units in the hidden layer;

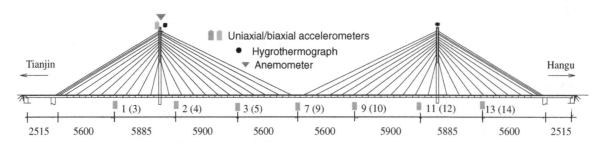

Figure 6. Scheme of the distribution of the sensors.

1, 3, 5, 7, 9 11, 13   2, 4, 6, 8, 10, 12, 14

Figure 7. Cross section of the bridge deck and position of the sensors.

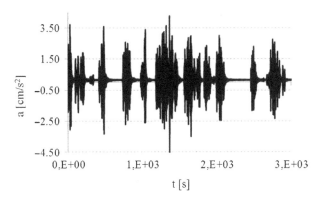

Figure 8. Time history of the acceleration at the station 2 on 17 January 2008.

the quantities within the brackets are known as *activations*, that are transformed using the activation functions h and g.

The values of the components of $\mathbf{w}$ are obtained during the training phase by minimising a proper error function that in the considered case is the sum of squared errors with weight decay regularisation (Bishop, 1995) given by:

$$E = \frac{1}{2}\sum_{n=1}^{N}\sum_{k=1}^{N_0}\left\{y_k(\mathbf{x}^n;\mathbf{w}) - t_k^n\right\}^2 + \frac{\alpha}{2}\sum_{i=1}^{W}|w_i|^2, \quad (2)$$

where y_k is the kth neural network output corresponding to the nth realisation of $\mathbf{x}$, t_k^n is the relevant target value, N is the size of the considered data set, N_0 is the number of output variables, W is the number of parameters in $\mathbf{w}$, α is a regularisation parameter.

The estimation of these parameters, i.e. the so-called model fitting, can be derived as a particular approximation of the Bayesian framework, and the entire neural network learning can be interpreted in the framework of Bayesian inference (MacKay, 1992), where probability is treated as a multi-valued logic that may be used to perform plausible inference (Jaynes, 2003). Starting from the early works of MacKay (1992) and Buntine and Weigend (1991), there has been a growing interest for the application of this framework in the field of neural networks methods (Bishop, 1995; Lampinen & Vethari, 2001; Nabney, 2004) also because within this framework it is possible to solve a crucial problem of neural network application: the choice of the optimal model architecture, i.e., the model with the

right complexity. In fact, for the neural network models, the number of adaptive parameters of the network model, i.e. the model class, has to be fixed in advance and affects significantly the generalisation performance of the network model. It is not correct to choose simply the model that fits the data better: more complex models will always fit the data better, but they may be over-parameterised and so they make poor predictions for new cases.

The problem of finding the optimal number of parameters provides an example of Ockham's razor, which states that best performance is achieved by the model whose complexity is neither too small nor too large. In general, the number of hidden units is selected by experience or rules of thumb and depends heavily on the subjective judgment of the designer; in this paper, the optimal architecture of the network model for a given set of training data is selected by a Bayesian model class selection approach (Beck & Yuen, 2004; Lam, Yuen, & Beck, 2006). As a result, the selection of the neural network model class is mathematically rigorous and systematic (MacKay, 1992). The most plausible model class among a set M of N_M candidate ones is obtained by applying Bayes' Theorem as follows:

$$p(M_j|D, M) \propto p(D|M_j)p(M_j|M). \quad (3)$$

The factor $p(D|M_jh)$ is known as the evidence for the model class M_j provided by the data D. Equation (3) shows that the most plausible model class is the one that maximises $p(D|M_j)p(M_j)$ with respect to j. If there is no particular reason *a priori* to prefer one model over another, they can be treated as equally plausible *a priori* and a non-informative prior, i.e. $p(M_j) = 1/N_M$, can be assigned; then different models can be compared just by evaluating their evidence (MacKay, 1992).

In the following, the optimal model has been chosen by comparing the evidence of different models with different numbers of adaptive parameters and selecting the model with the highest value of evidence.

Damage detection strategy

The proposed strategy is an extension of the approach that was proposed by Arangio and Beck (2012) on simulated data. For the first time, it was tested on real noisy data. It consists in building a system of neural networks working in parallel, which are able to represent and approximate the behaviour of the bridge in undamaged conditions. For this purpose, different neural network models, one for each measurement location and one for each hour of measurements [that is, the number of network models is 14 (locations $\times$ 24 (h) = 336)], were built. The neural network models are trained using the time histories of the accelerations recorded in the selected locations in the undamaged condition. In this way, the models are able to

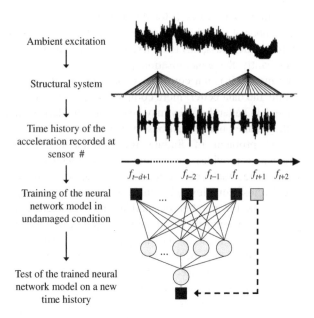

Figure 9.   Scheme of the proposed damage detection strategy.

approximate the behaviour of the undamaged bridge taking into account the different use during the different hours of the day.

The procedure for network training is shown in Figure 9. The time history of the response f is sampled at regular intervals, generating series of discrete values f_t. For this purpose, the time series presented in Section 1.2 were decimated, by a factor of 10. Moreover, in order to obtain signals that could be adequately reproduced, the time series was pre-processed by scaling the data within the nonlinear range of the activation function, where it is possible to avoid the saturation at the asymptotes of the activation function. After that, a set d of values of the processed time series, $f_{t-d+1}, \cdots, f_t$, is used as input of the network model, while the next value f_{t+1} is used as target output. By stepping along the time axis, a training data set consisting of many sets of input vectors with the corresponding output values is built, and the network models are trained.

The performance of the trained models is tested by proposing to the network input patterns of values recorded some minutes after those used for training $f_{t+n-d}, \cdots, f_{t+n}$, and by predicting the value of f_{t+n+1}. The models were considered well trained because they showed to be able to reproduce the expected values with a small error. Then, these trained neural networks models are tested with data recorded in the following days. The testing patterns correspond to both undamaged and damaged conditions. For each pattern, the next value is predicted and compared with the target output. If the error in the prediction is negligible, the models show to be able to reproduce the monitoring data and the bridge is considered undamaged; if the error in one or more locations is large, the presence

of an anomaly (that may represent or may not represent damage) is detected.

Optimal model class selection

For the selection of the optimal model, the Bayesian approach has been applied. A class of models with four input units, corresponding to the number of previous instants needed to predict the value at the instant $t + 1$, was considered. The optimal number of input units was obtained by applying the so-called automatic relevance determination approach. More details on the method are given in Arangio and Beck (2012). For this class with four inputs, different architectures, with an increasing number of hidden units $(1, 2, \ldots)$, were analysed and compared. As already said, all the models can be treated as equally plausible a priori and a non-informative prior $p(M_j) = 1/N_M$ over the models can be assigned. In this way, the various models were compared by just evaluating their evidence $p(D|M_j)$ that was computed using the following relation (Equation (3)) based on Laplace's asymptotic approximation of the evidence integral (Bishop, 1995):

$$\ln p(D|M_j) = -\beta_j^{MP} E_D^{MP} + \frac{N}{2} \ln \beta_j^{MP} + \ln (H_j^{MP}!)$$
$$+ 2 \ln H_j^{MP},$$

$$(4-\text{part}\,1)$$

$$-\alpha_j^{MP} E_W^{MP} - \frac{1}{2} \ln \left| \mathbf{A}_j^{MP} \right| + \frac{W_j}{2} \ln \alpha_j^{MP}$$
$$+ \frac{1}{2} \ln \left(\frac{2}{\gamma_j^{MP}} \right) + \frac{1}{2} \ln \left(\frac{2}{N - \gamma_j^{MP}} \right) H_j^{MP}.$$

$$(4-\text{part}\,2)$$

The detailed explanation of the meaning of each term of the equation is beyond the scope of this work. In order to give an idea of the involved parameters that will be recalled in the following, α_j and β_j are the maximum a posteriori values of the hyper parameters for model class M_j (parameters related to the model fitting), γ_j is a parameter that can be considered a measure of the well-determined parameters, N is the number of samples in the data set, H_j^{MP} is the number of hidden units of the model class M_j, the matrix A_j is the Hessian of the total error function (composed by two different error terms E_D and E_W), and W_j is the number of parameters in $\mathbf{w_j}$. More details about the mathematical aspects of the approach are given in Bishop (1995).

The mathematical relationship has been written for pointing out the fact that in this equation, it is possible to recognise two kinds of terms. Those in Equation ((4) – part 1) give a measure of the data fit, so they increase as the models get more complex. The sum of the other terms (Equation (4) – part 2), which represents the so-called Ockham factor, penalises more complex models, so it

Table 1. Bayesian optimisation of the neural network model.

Model	1	2	3	4	5
N parameters	7	13	19	25	31
Gamma	2.00	3.03	4.02	5.00	6.00
$-\beta_j^{MP} E_D^{MP} + (N/2) \ln \beta_j^{MP}$	20,770	22,682	25,078	22,153	23,500
$\ln \left(H_j^{MP}! \right) + 2 \ln H_j^{MP}$	2.08	3.99	5.95	8.01	10.16
Data fit term	20,772	22,686	25,084	22,161	23,510
$-\alpha_j^{MP} E_W^{MP} + (1/2) \ln |\mathbf{A}| + (W/2) \ln \alpha_j^{MP}$	−13.08	−79.32	−158	−213	−266
$1/2 \ln (2/\gamma) + (1/2) \ln (2/N - \gamma)$	−3.31	−3.51	−3.66	−3.8	−3.86
Penalising term	−16	−83	−162	−217	−270
Log evidence	20,756	22,603	24,922	21,944	23,240

decreases when the complexity increases. This sum can be viewed as the amount of information extracted from the data by model class M_j (Beck & Yuen, 2004). The optimal complexity is given by the model with the best compromise between these terms. In Table 1, the logarithm of the evidence has been computed for various models at increasing complexity (that is with more hidden units) within the class with four input variables. The terms in the light grey boxes are related to the data fit, so they tend to increase with the complexity, while the terms in the deep grey boxes penalise too complex models, so they tend to decrease as the models become more complex. The model with the right compromise is the one with the highest evidence, which results to be the one with three hidden units. Therefore, four-three-one is the architecture chosen for all the networks models.

The results of the training and test phases are reported in Figure 10. The quantities measured by the sensors were

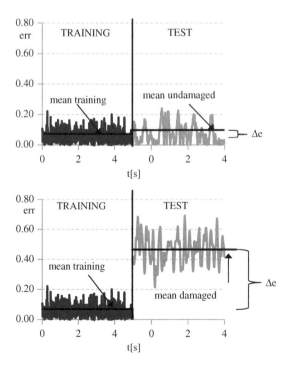

Figure 10. Error in the approximation for training and test in health and damaged conditions.

the acceleration $a(t)$. After the pre-processing they become the input and the target output of the network model, called respectively x and t, while it will be y is the output given by the network model. The two plots in Figure 10 show the difference between the obtained network output value y and the expected target value t at several time steps for both training and testing, in undamaged and damaged conditions. This quantity is indicated as err. If the structure remains undamaged, the difference between the two values of err, called Δe, is close to 0. On the contrary, in case of anomalies that may correspond to damage, there is a significant difference Δe between the mean values of error in testing and training.

To distinguish the actual cause of the anomaly, the intensity of Δe is observed at different measurement locations: if Δe is large in several locations, it can be concluded that the external actions (wind, traffic) are probably changed. In this case, the trained neural network models are unable to represent the time histories of the response parameters, and they have to be updated and retrained according to the modified characteristics of the action. If Δe is large only in one or few locations, it can be concluded that the bridge experienced some damages.

Results of the proposed strategy

In the sequence, the results of the damage detection strategy applied to the ANCRiSST benchmark problem are shown. As said above, 14 groups of neural networks have been created, one group for each measurement location, which have been trained with the time histories of the accelerations in health conditions (data recorded on 17 January 2008). In order to take into account the change in the vibrations of the structures caused by the different use during the day, one network model for each hour of monitoring has been created (24 network models for each location).

For the training phase of each model, four steps of the considered time history are given as input and the following step as output. As said above, the needed time steps to be used as input have been determined by using the Automatic Relevance Determination (Arangio & Beck, 2012). The training set of each network model

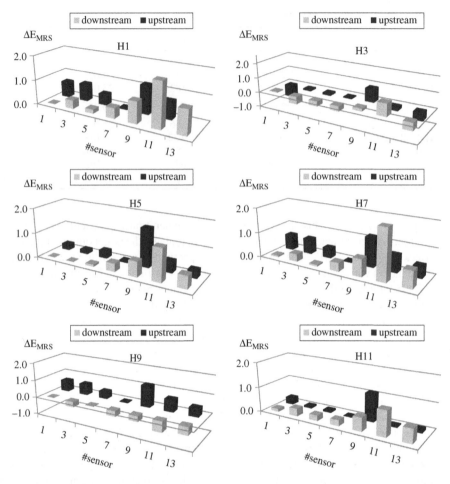

Figure 11. Differences of the root mean square of the error ΔE_{RMS} in the 14 locations of the sensors (from H1 to H11).

includes 5000 examples chosen randomly in the entire set. The trained networks have been tested by using the time histories of the accelerations recorded at the same locations and at the same time on 31 July 2008. In order to evaluate the error, it was used the root mean squares of the error, E_{RMS}, given by:

$$E_{RMS} = \sqrt{\frac{2E(\mathbf{w})}{N}},$$

where E is the sum of squares error in Equation (2), $\mathbf{w}$ is the weight vector of the trained network, N is the size of the data set.

E_{RMS} was calculated for the time series recorded in both considered dates (January 17 and July 31), and Figures 11 and 12 present the difference between these two values, called ΔE_{RMS}. Each plot shows the results of 1 of the 24 h of the day (called as H1, H3, etc.). The results every 2 h are shown. Each graph has 14 bars and each bar represents the value of ΔE_{RMS} for one of the 14 sensors. The bars in the plot are arranged in the same way of the sensors along the deck of the bridge: the seven bars in light grey represent the results of the downstream sensor (#1, 3,

5, 7, 9, 11, 13 in Figures 6 and 7), while the bars in dark grey represent those of the upstream sensors (#2, 4, 6, 8, 9, 10, 12, 14 in Figures 6 and 7).

By observing the plots, it is easy to notice that, apart some hours of the day that look difficult to reproduce, the neural networks models are able to approximate the time history of the acceleration with a small error in almost all the measurement locations, except that around sensor #10. Considering that in the undamaged condition the error was small in all the locations, this difference is interpreted as the presence of an anomaly (damage) in the structure. Between 6 a.m. and 9 a.m. and around 9 p.m., the error is larger in various sensors, but it is possible that this depends on the additional vibrations given by the traffic in the busiest hours of operation of the bridge. This aspect will be investigated in order to introduce improvements in the methodology that allow a fruitful application of the method in all the time of the day. Notice that the results in Figures 11 and 12 represent a first result, but the detected anomaly would need further investigation; data obtained by the additional field testing carried out on August 2008 could be used to develop a strategy for the localisation of the damage.

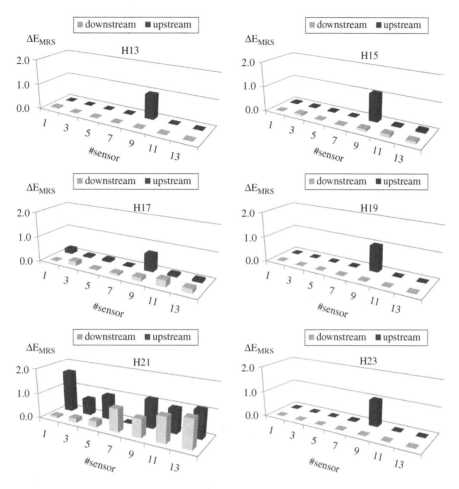

Figure 12.   Differences of the root mean square of the error ΔE_{RMS} in the 14 locations of the sensors (from H13 to H23).

Suggestion of damage presence by traditional structural identification

As seen earlier, the proposed neural networks-based approach detected the presence of anomalies in the behaviour of the structure that could be related to a degradation or a damage of the bridge. In order to validate this decrease of the performance of the structure, the time series given by the accelerometers were analysed also by carrying out a traditional vibration-based structural identification. More details are given in Mannucci (2012).

Generally, the structural identification of a civil structure includes the evaluation of its modal parameters, which are able to describe the dynamic behaviour of the entire structure. During the last three decades, extensive research has been conducted in vibration-based identification methods and significant progress has been achieved (Doebling et al., 1996; Ewins, 2000; Gul & Catbas, 2008; Sohn et al. 2004). In this work, the identification was carried out by using the enhanced frequency domain decomposition (EFDD) technique (Brincker, Zhang, & Andersen, 2001). With the frequency domain decompo-

sition (FDD) technique, the frequency content of the response is analysed by using the auto-cross power spectral density (PSD) functions of the measured time series of the responses. The PSD matrix is then decomposed by using the singular value decomposition (SVD) tool. The singular values (SV) contain information from all spectral density functions. The peaks of a singular value plot indicate the existence of different structural modes, so they can be interpreted as the auto-spectral densities of the modal coordinates, and the singular vectors as mode shapes (Brincker et al., 2001).

It should be noted that this approach is exact when the considered structure is lightly damped and excited by a white noise, and when the mode shapes of closed modes are geometrically orthogonal (Ewins, 2000). If these assumptions are not completely satisfied, the SVD is an approximation, but the obtained modal information is still enough accurate (Brincker, Ventura, & Andersen, 2003). The first step of the FDD is to construct a PSD matrix of ambient responses $G(f)$ that, in the considered case, was computed by using the Welch's averaged modified

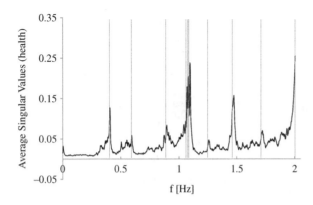

Figure 13. Averaged SV decomposition (health condition).

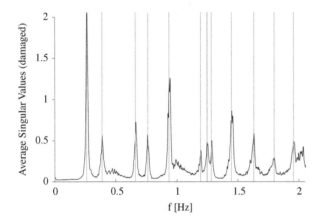

Figure 14. Averaged SV decomposition (damaged condition).

periodogram method (Welch, 1967). In addition, an overlapping of 50% between the various segments was considered, and a periodic Hamming windowing was applied to reduce the leakage. After the evaluation of the spectral matrix, the FDD technique involves the SVD of G (f) at each frequency and the inspection of the curves representing the SV.

The SVD have been carried out for all the time series of the accelerations recorder over the 24 h on 17 January 2008. The averaged SVD plot is shown in Figure 13. The attention was focused on the frequencies below 2 Hz. The selection of this range has been done for two reasons: first, because the most important modes for the dynamic description of large structural systems generally are below 2 Hz; in addition, the available data included the measurements of 14 stations (seven downstream and seven upstream) that make difficult to identify clearly higher frequency. The same procedure has been applied for processing the time series of the response in damaged conditions. In Figure 14, the averaged SVD is shown. It is possible to note three SV coming up around 1.1 and 1.3 Hz that indicate the presence of three modes in this range. The other modes are reasonably separated.

Table 2. Comparison of the frequencies of the first six modes obtained from the FEM and from the vibration-based identification in undamaged and damaged conditions.

Mode	Frequencies		
	f_{FEM}	f_{undam}	f_{dam}
1	0.452	0.4075	0.262
2	0.632	0.594	0.388
3	0.937	0.896	0.664
4	1.106	1.096	0.76
5	1.270	1.262	0.94
6	1.503	1.472	1.194

The results of the vibration-based identification have been compared with the output of the modal analysis carried out with the FEM of the structure (shown in Figure 3). For this comparison, it should be considered that the FEM represents the 'as built' bridge where the mechanical properties and the cross sections were assigned as reported in the original project, while the recorded accelerations represent the behaviour of the bridge after years of operation. The comparison of the first six frequencies is summarised in Table 2, and the first three mode shapes are shown in Figure 15.

By inspecting this figure, it can be noticed that the mode shapes identified using the time series recorded in undamaged condition are in good agreement with those given by the FEM. The mode shapes remain similar also after damage because probably it affects the higher modes. However, the decrement of the frequencies implies the deterioration of the structure or the occurrence of damage. The frequencies of the FEM model, which represent the 'as built' structure, are higher of those obtained from the signal recorded in January 2008, showing that the years of operation have reduced the overall stiffness of the structure. This phenomenon is even more evident looking at the decrement of the frequencies of the bridge in the damaged condition.

Note that there is another factor that was not object of this study but that could have partially influenced the variation of the frequencies, as stated by Li et al. (2010): the dependence on the temperature. Actually, the two signals have been recorded in two different periods of the year that are characterised by a significant difference of temperature. However, the results are in substantial accordance with those obtained with the neural network model because they suggest that the decrement of the frequencies does not depend only on the temperature, but it could be related to the presence of deterioration or damage. So, both methods, in completely different ways, suggest that the overall behaviour of the bridge has changed, and it requires further attention of the Owner Authority.

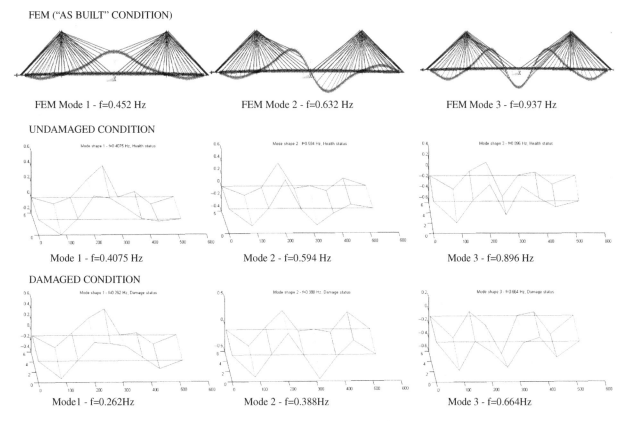

FEM ("AS BUILT" CONDITION)

FEM Mode 1 - f=0.452 Hz   FEM Mode 2 - f=0.632 Hz   FEM Mode 3 - f=0.937 Hz

UNDAMAGED CONDITION

Mode 1 - f=0.4075 Hz   Mode 2 - f=0.594 Hz   Mode 3 - f=0.896 Hz

DAMAGED CONDITION

Mode1 - f=0.262Hz   Mode 2 - f=0.388Hz   Mode 3 - f=0.664Hz

Figure 15. Comparison of the first three mode shapes obtained from the FEM and from the vibration-based identification in undamaged and damaged conditions.

Conclusions

This work presents the results of a neural networks-based strategy for damage detection that has been tested on a benchmark problem proposed by the Asian-Pacific Network of Centers for Research in Smart Structure Technology (ANCRiSST). The ANCRiSST opened to the researchers the vibration data obtained from the continuous monitoring of a cable-stayed bridge in Mainland China. The available data set includes time histories of uniaxial accelerations in 14 different locations in two different dates (17 January and 31 July 2008). In the timespan between the two available registrations, the bridge experienced two damages.

The scope of the proposed strategy is the detection of damage by using only the available time series of the registration. For this purpose, various neural network models have been developed. The optimal architecture of the models has been chosen by applying the Bayesian model class selection. The various models have been trained to approximate the time histories of the structural response in the health condition. Subsequently, new inputs, corresponding to both health and damaged conditions, have been tested on the trained models. Analysing the increment of the error in the prediction, the proposed approach suggests the presence of an anomaly in the behaviour of the bridge. The obtained results have been basically confirmed by applying a traditional vibration-based structural identification technique, the EFDD. The modal properties of the structure have been obtained on the base of the same data set used with the Bayesian neural networks, and the frequencies and mode shapes in undamaged and damaged conditions have been compared, showing a remarkable decrement of the frequencies that could be symptoms of an anomaly or a damage.

Acknowledgements

Prof. Hui Li and Wensong Zhou of the Harbin Institute of Technology, Eng. Silvia Mannucci, the team of Prof. Bontempi's from Sapienza University of Rome are gratefully acknowledged. Prof. Jim Beck of Caltech is acknowledged for his contribution at the initial development of the strategy. This research was partially supported by StroNGER s.r.l. from the fund 'FILAS – POR FESR LAZIO 2007/2013 – Support for the research spin off'. However, the opinions and conclusions presented in this paper are those of the authors and do not necessarily reflect the views of the sponsoring organisations.

Note

1. Email: franco.bontempi@uniroma1.it

References

Adeli, H. (2001). Neural networks in civil engineering: 1989–2000. *Computer-Aided Civil and Infrastructure Engineering*, *16*, 126–142.

ANCRiSST. (2011). *ANCRiSST SHM benchmark problem*. Harbin: Center of Structural Monitoring and Control of the Harbin Institute of Technology. http://smc.hit.edu.cn/index.php?option=com_content&view=article&id=121&Itemid=81

Arangio, S. (2012). Reliability based approach for structural design and assessment: Performance criteria and indicators in current European codes and guidelines. *International Journal of Lifecycle Performance Engineering*, *1*, 64–91.

Arangio, S., & Beck, J.L. (2012). Bayesian neural networks for bridges integrity assessment. *Structural Control & Health Monitoring*, *19*, 3–21.

Arangio, S., & Bontempi, F. (2010). Soft computing based multilevel strategy for bridge integrity monitoring. *Computer-Aided Civil and Infrastructure Engineering*, *25*, 348–362.

Arangio, S., & Bontempi, F. (2014). Design knowledge gain by structural health monitoring. In D.M. Frangopol & Y. Tsompanakis (Eds.), *Safety and maintenance of aging infrastructure* (Chap. 4, pp. 95–120). Boca Raton, FL: CRC Press.

Arangio, S., Calò, F., Di Mauro, M., Bonano, M., Marsella, M., & Manunta, M. (2014). An application of the SBAS-DInSAR technique for the assessment of structural damage in the city of Rome (Italy). *Structure and Infrastructure Engineering*. doi:10.1080/15732479.2013.833949

Beck, J.L., & Yuen, K.-V. (2004). Model selection using response measurements: Bayesian probabilistic approach. *Journal of Engineering Mechanics*, *130*, 192–203.

Biondini, F., Frangopol, D.M., & Malerba, P.G. (2008). Uncertainty effects on lifetime structural performance of cable-stayed bridges. *Probabilistic Engineering Mechanics*, *23*, 509–522.

Bishop, C.M. (1995). *Neural networks for pattern recognition*. Oxford: Oxford University Press.

Bontempi, F., & Giuliani, L. (2010). Basic aspects for the uncertainty in the design and analysis of bridges. In Dan Frangopol, Richard Sause, & Chad Kusko (Eds.), *5th international conference on bridge maintenance, safety and management* (IABMAS 2010) (pp. 2205–2212). Philadelphia, PA, 11–15 July. Leiden, The Netherlands: CRC Press.

Bontempi, F., Gkoumas, K., & Arangio, S. (2008). Systemic approach for the maintenance of complex structural systems. *Structure and Infrastructure Engineering*, *4*, 77–94.

Brincker, R., Ventura, C. E., Andersen, P. (2003). Why output-only modal testing is a desirable tool for a wide range of practical applications. *21st international modal analysis conference (IMAC-XXI)*, (pp. 8–16), Kissimmee, FL, 3–6 February.

Brincker, L., Zhang, L., & Andersen, P. (2001). Modal identification of output-only systems using frequency domain decomposition. *Smart Materials and Structures*, *10*, 441–445.

Buntine, W., & Weigend, A. (1991). Bayesian back-propagation. *Complex System*, *5*, 603–643.

Casas, J.R. (2010). Assessment and monitoring of existing bridges to avoid unnecessary strengthening or replacement. In Dan Frangopol, Richard Sause, & Chad Kusko (Eds.), *5th international conference on bridge maintenance, safety and management* (IABMAS 2010) (pp. 2268–2276). Philadelphia, PA, 11–15 July. Leiden, The Netherlands: CRC Press.

Ceravolo, R., De Stefano, A., & Sabia, D. (1995). Hierarchical use of neural techniques in structural damage recognition. *Smart Materials and Structures*, *4*, 270–280.

Choo, J.F., & Koh, H.M. (2009). A neural network-based damage detection algorithm using dynamic responses measured in civil structures. In *Fifth international joint conference on INC, IMS and IDC 2009* (pp. 682–685). Seoul. IEEE.

Crosti, C., Duthinh, D., & Simiu, E. (2011). Risk consistency and synergy in multihazard design. *Journal of Structural Engineering - ASCE*, *137*, 844–849.

Doebling, S.W., Farrar, C.R., Prime, M.B., & Shevitz, D.W. (1996). *Damage identification and health monitoring of structural and mechanical systems from changes in their vibration characteristics: A literature review*. Los Alamos National Laboratory Report LA-13070-MS 1996. Los Alamos, NM: Los Alamos National Laboratory.

Dordoni, S., Malerba, P.G., Sgambi, L., & Manenti, S. (2010). Fuzzy reliability assessment of bridge piers in presence of scouring. In Dan Frangopol, Richard Sause, & Chad Kusko (Eds.), *5th international conference on bridge maintenance, safety and management* (IABMAS 2010) (pp. 1388–1395). Philadelphia, PA, 11–15 July. Leiden, The Netherlands: CRC Press.

Ewins, D.J. (2000). *Modal testing: Theory, practice and application* (2nd ed.). Baldock: Research Studies Press.

Frangopol, D.M., Saydam, D., & Kim, S. (2012). Maintenance, management, life-cycle design and performance of structures and infrastructures: a brief review. *Structure and Infrastructure Engineering*, *8*(1), 1–25.

Freitag, S., Graf, W., & Kaliske, M. (2011). Recurrent neural networks for fuzzy data. *Integrated Computer-Aided Engineering – Data Mining in Engineering*, *18*, 265–280.

Gul, M., & Catbas, F.N. (2008). Ambient vibration data analysis for structural identification and global condition assessment. *Journal of Engineering Mechanics*, *134*, 650–662.

Jaynes, E.T. (2003). *Probability theory: The logic of science*. Cambridge: Cambridge University Press.

Kim, S.H., Yoon, C., & Kim, B.J. (2000). Structural monitoring system based on sensitivity analysis and a neural network. *Computer-Aided Civil and Infrastructure Engineering*, *15*, 309–318.

Ko, J.M., Ni, Y.Q., Zhou, H.F., Wang, J.Y., & Zhou, X.T. (2009). Investigation concerning structural health monitoring of an instrumented cable-stayed bridge. *Structure and Infrastructure Engineering*, *5*, 497–513.

Ko, J.M., Sun, Z.G., & Ni, Y.Q. (2002). Multi-stage identification scheme for detecting damage in cable-stayed Kap Shui Mun Bridge. *Engineering Structures*, *24*, 857–868.

Koh, H.M., Kim, H.J., Lim, J.H., Kang, S.C., & Choo, J.F. (2010). Lifetime design of cable-supported super-long-span bridges. In Dan Frangopol, Richard Sause, & Chad Kusko (Eds.), *5th international conference on bridge maintenance, safety and management* (IABMAS 2010) (pp. 35–52). Philadelphia, PA, 11–15 July. Leiden, The Netherlands: CRC Press.

Lam, H.F., Yuen, K.V., & Beck, J.L. (2006). Structural health monitoring via measured Ritz vectors utilizing artificial neural networks. *Computer-Aided Civil and Infrastructure Engineering*, *21*, 232–241.

Lampinen, J., & Vethari, A. (2001). Bayesian approach for neural networks: Review and case studies. *Neural Networks*, *14*, 257–274.

Lan, C., Zhou, Z., Sun, S., & Ou, J. (2008). FBG based intelligent monitoring system of the Tianjin Yonghe Bridge. *Proceedings SPIE 6933, Smart Sensor Phenomena,*

Technology, Networks, and Systems 2008, 693312. doi:10.1117/12.777267

Lapedes, A., & Farber, R. (1987). *Non linear signal processing using neural networks*. Los Alamos National Laboratory Technical Report LA-UR-87-2662. Los Alamos, NM: Los Alamos National Laboratory.

Li, H., Li, S., Ou, J., & Li, H. (2010). Modal identification of bridges under varying environmental conditions: temperature and wind effects. *Structural Control and Health Monitoring, 17*, 495–512.

Li, H., Ou, J., Zhao, X., Zhou, W., Li, H., & Zhou, Z. (2006). Structural health monitoring system for Shandong Binzhou Yellow River Highway Bridge. *Computer-Aided Civil and Infrastructure Engineering, 21*, 306–317.

MacKay, D.J.C. (1992). A practical Bayesian framework for back-propagation networks. *Neural Computation, 4*, 448–472.

Mannucci, S. (2012). Enhanced structural identification of Yonghe Bridge (Master thesis in Civil Engineering). Sapienza University of Rome, Italy.

Nabney, I.T. (2004). *NETLAB: Algorithms for pattern recognition*. New York, NY: Springer.

Ni, Y.Q., Wong, B.S., & Ko, J.M. (2002). Constructing input vectors to neural networks for structural damage identification. *Smart Materials and Structures, 11*, 825–833.

Olmati, P., & Gentili, F. (2010). Structural response of bridges to fire after explosion. In Dan Frangopol, Richard Sause, & Chad Kusko (Eds.), *6th international conference on bridge maintenance, safety and management* (IABMAS 2012) (pp. 2017–2023). Stresa, Lake Maggiore, 8–12 July. Leiden, The Netherlands: CRC Press.

Petrini, F., & Bontempi, F. (2011). Estimation of fatigue life for long span suspension bridge hangers under wind action and train transit. *Structure and Infrastructure Engineering, 7*, 491–507.

Qui, M., & Zhang, G.P. (2003). Time series modeling and forecasting with neural networks. *Proceedings of the international conference on computational intelligence for financial engineering*, 20–23 March, CIFE03, Hong Kong.

Sgambi, L., Gkoumas, K., & Bontempi, F. (2012). Genetic algorithms for the dependability assurance in the design of a long-span suspension bridge. *Computer-Aided Civil and Infrastructure Engineering, 27*, 655–675.

Sohn, H., Farrar, C.R., Hemez, F.M., Shunk, D.D., Stinemates, D.W., Nadler, B.R., & Czarnecki, J.J. (2004). A review of structural health monitoring literature: 1996–2001. *Los Alamos National Laboratory Report LA-13976-MS 2004*. Los Alamos, NM: Los Alamos National Laboratory.

Tsompanakis, Y., Lagaros, N.D., & Stavroulakis, G. (2008). Soft computing techniques in parameter identification and probabilistic seismic analysis of structures. *Advances in Engineering Software, 39*, 612–624.

Waszczyszyn, Z., & Ziemianski, L. (1999). Neural networks in the identification analysis of structural mechanics problems. In Z. Mroz & G.E. Stavroulakis (Eds.), *Parameter identification of materials and structures*. New York, NY: Springer.

Welch, D. (1967). The use of fast Fourier transform for the estimation of power spectra: A method based on time averaging over short modified periodograms. *IEEE Trans Audio Electroacoustic, 15*, 70–73.

Xu, H., & Humar, J. (2006). Damage detection in a girder bridge by artificial neural network technique. *Computer-Aided Civil and Infrastructure Engineering, 21*, 450–464.

Real-time remote monitoring: the DuraMote platform and experiments towards future, advanced, large-scale SCADA systems

Masanobu Shinozuka, Konstantinos G. Papakonstantinou, Marco Torbol and Sehwan Kim

The economic and social prosperity of our society depends much on the proper functioning of structures and civil infrastructure systems. Structural health monitoring has been long recognised as a vital tool to preclude and/or mitigate degradation effects and failures of structural systems. Along this line, the DuraMote platform is presented in this paper (named after Durable and Mote) together with real-life applications, laboratory and field experiments, which promote the effort to expand the existing concept of structural monitoring into remote, real-time, continuous and permanent performance monitoring of spatially extended systems. Successful implementation of this technology can improve the resilience and sustainability of large-scale complex infrastructure systems and lead to future, advanced Supervisory Control and Data Acquisition methodology that could be routinely used in a variety of structures and networks.

Nomenclature

10 Base-T:	10 Mbps twisted pair ethernet
100 Base-FX:	100 Mbps two pairs fiber-optic cables
ADC:	analogue to digital converter
AP:	access point
BPSK:	binary phase shift keying
CAN:	controller area network
DuraMote:	durable sensing mote
FHSS:	frequency hopping spread spectrum
FIR:	finite impulse response
FRAM:	ferroelectric random access memory
GFSK:	Gaussian frequency shift keying
GPS:	global positioning system
I^2C:	inter-integrated circuit
LAN:	local area network
LOS:	line of sight
MCU:	microcontroller unit
MEMS:	microelectromechanical system
MPPT:	maximum power point tracking
OFDM:	orthogonal frequency division multiplexing
P to MP:	point to multi-point
P to P:	point to point
PCB:	printed circuit board
PHY:	physical layer
PoCAN:	power over controller area network
QAM:	quadrature amplitude modulation
QPSK:	quadrature phase shift keying
RF:	radio frequency
RTC:	real time clock
RX:	reception
SCADA:	supervisory control and data acquisition
SDHC:	secure digital high capacity
SHM:	structural health monitoring
SPI:	serial peripheral interface
TDMA:	time division multiple access
TX:	transmission
WDS:	wireless distribution system
Wi-Fi:	wireless fidelity
WWVB:	standard time and frequency station, 60 KHz, Fort Collins, Colorado

1. Introduction

The economic and social prosperity of our society depends much on the proper functioning of structures and civil infrastructure systems. Structural health monitoring has been long recognised as a vital tool to preclude and/or mitigate degradation effects and failures of structural systems. In Ko and Ni (2005), different wired structural health monitoring systems and key technology issues for large-scale bridges are outlined, together with some insights about the linkage between structural health monitoring and bridge inspection, maintenance and management. Some fundamentals about distributed wireless sensor networks can be read in Lewis (2004). Development of wireless sensor units and networks for structural monitoring can be seen in Lynch et al. (2003) and Xu et al. (2004), with a summary review available in Lynch and Loh (2006), which is also useful in order to compare important features between our DuraMote platform, presented in this paper, and other wireless sensing systems. Further approaches and trends can be also read in Rice et al. (2010), Rice, Mechitov, Sim, Spencer, and Agha (2011), while Pakzad, Fenves, Kim, and Culler (2008) tested a wireless sensor network on the Golden Gate Bridge in San Francisco to test its multi-hop communications, scalability and performance. The important and practical subject of energy harvesting for structural health monitoring sensor networks is analysed in Park, Rosing, Todd, Farrar, and Hodgkiss (2008). Ho et al. (2012) deployed a solar-powered vibration and impedance sensor system, based on the iMote2, on the cable-stayed Hwamyung Bridge in South Korea. The solar panel was directly connected to a battery through a diode without performing maximum power point tracking (MPPT) though.

The approaches presented so far mainly concern single structures and not spatially distributed large scale civil infrastructure systems. In addition, they mostly focus on event-driven and not on continuous data streaming types of applications. Permanent, continuous monitoring of spatially distributed systems is a more difficult and demanding problem that can provide immense societal benefits, both in economic and human welfare terms, if successfully tackled. Among the myriads of situations where this technology can be of significance, energy and water pipeline systems are mentioned in this work, with emphasis on the latter.

In general, breaks and leakages in pipeline systems result in significant economic losses, and disastrous societal disruption. In terms of economic losses, the economic costs associated with breaks of deteriorating pipelines are constantly rising. A recent report of the US Environmental Protection Agency (EPA, 2010) claims that there are 240,000 water main breaks per year in the USA and the US Geological Survey estimates that water lost from water distribution systems is 1.7 trillion gallons per year at a national annual cost of $2.6 billion. Although it is thus widely recognised that pipelines sustain damage quite often, the current technology falls short in handling the problem by detecting locations of breaks/leakages, immediately respond and effectively mitigate the damage. Based on data from the Pipeline and Hazardous Materials Safety Administration collected between 2002 and 2012, (Song, 2012), leak detection systems and controllers in control rooms identified only 5% and 7%, respectively, of

oil pipeline spills, out of 960 total leaks, and 20% and 18%, respectively, out of 71 significant spills over 1000 barrels. Yet, in the second case, the general public discovered almost as many spills (17%) and 42% were discovered by employees at the scenes of accidents. Another indicative example about the capabilities of current technology is the occurred pipeline rupture near Marshall, Michigan on 25 July 2010. It took operators in owner's control room several hours to realise and verify that the pipeline had torn open and before the spill was contained around one million gallons of diluted bitumen escaped from the tear along the seam of the pipe into the Kalamazoo River and surrounding wetlands.

Continuous monitoring of such systems has been long thought as a solution for the purpose of system operation. However, current Supervisory Control and Data Acquisition (SCADA) systems are not designed for damage localisation. Conventional SCADA systems, widely deployed in utility industry, sense and control operational perturbations in pressure, flow rate, temperature, chemical quality data, etc. Sensors are installed at key components, such as pump stations, but usually not along pipe networks. Current SCADA systems can in principle be used for damage-detection purposes. However, the usual sparseness of the data acquisition locations within the network makes it difficult to detect the damage with accuracy and rapidity in practice. For a concise review of different leak detection methods, based on SCADA systems or not, the interested reader is directed to Zhang (1996) and Colombo, Lee, and Karney (2009).

Towards the development of a future, advanced SCADA system, equipped with a dense wireless sensor network along the entire pipeline length, a series of publications have been lately written. Usually, the sensor modules consist of a variety of sensors, such as pressure gauges, hydrophones, flow meters, PH sensors, etc. Next generation SCADA approaches vary in their sensing techniques, mathematical formulation, data acquisition methods, and data processing algorithms, among others. In Stoianov, Nachman, Madden, and Tokmouline (2007) PipeNet is presented, which is a prototype tiered wireless monitoring system for hydraulic and water quality parameters. PipeNet has three tiers: Mote tier in manholes, Gateway tier placed on the utility poles, and Middleware and Backend tier at a central office. The Stargate in the Gateway tier consumes high computational and long-range wireless power. Thus, it is powered by a grid with a battery backup. The Mote tier consists of a cluster of battery-operated sensor motes, with low data storage, performing signal compression and local data processing. In Jin and Eydgahi (2008), non-destructive detection technology is utilised, based on Piezoelectric ceramic lead Zirconate Titanate sensors, which are mounted on the curved surface of the pipeline for generating and measuring guided waves along the pipes. Kim, Schmid, Charbiwala, Friedman, and Srivastava (2008) introduce a two-tier non-intrusive water monitoring system, NAWMS, which captures the water flow rate using vibration sensors and uses similar hardware components as PipeNet. Yoon, Ye, Heidemann, Littlefield, and Shahabi (2011) suggest a Steamflood and Waterflood Tracking System for oil field monitoring, improving the conventional SCADA systems by adopting wireless sensor networks with a decentralised architecture. Finally, WaterWise is an ongoing effort for continuous remote monitoring of water distribution systems. WaterWise represents advancements beyond the accomplishments of PipeNet and facilitates in situ experimentation and development since it currently monitors a $60\,\mathrm{km}^2$ area of downtown Singapore (Allen et al., 2011).

In this work, the latest version of our developed DuraMote platform is presented that is a new cost-effective, advanced remote monitoring and inspection system. DuraMote takes its name from Durable and Mote, which is often used as a suffix indicating small-size sensor nodes. It is the latest outcome of efforts by Shinozuka and co-workers regarding remote monitoring systems and the evolution of the earlier platforms, DuraNode, (Shinozuka, Park, Chou, & Fukuda, 2006) and PipeTECT (Kim, Chou, & Shinozuka, 2011; Shinozuka, Chou, Kim, Kim, Yoon et al., 2010). DuraMote is designed as a very flexible remote monitoring system that can address the requirements of a diverse and wide range of structural health monitoring applications, from single structures to next generation, large-scale, spatially distributed SCADA systems. It adopts a tiered configuration and consists of data aggregators named Roocas, and sensing nodes named Gophers. It supports both wireless and wired communication capabilities and it successfully addresses a number of challenges, such as the ability to cover a large area, to handle multiple, different sensor units, low power demand for sensing and transmission, continuous monitoring and so on. According to application needs, DuraMote can be straightforwardly adjusted with appropriate sensors, sampling rate, cut-off frequency, communication throughput and interface, power scheme and so forth. Energy harvesting capabilities with MPPT and supercapacitors are feasible as well based on works by Kim and Chou (2011) and Kim, No, and Chou (2011). Even for SCADA applications, DuraMote's sensing node is mainly based on MEMS accelerometers, unlike other SCADA approaches previously mentioned. The reason for this is that a new non-invasive method for damage detection of pipes is utilised that relies on vibrations of the pipe walls, (Papakonstantinou, Shinozuka, & Beikae, 2011). More details on this methodology will be presented later on in the paper.

This work aims to promote the effort to expand the existing concept of structural monitoring into remote, real-time, continuous and permanent performance monitoring of spatially extended systems and structures. To this end,

in the rest of the paper, the DuraMote platform is presented in more detail, together with a variety of laboratory and field experiments on long-span bridges, buildings and pipeline networks. A consolidated terminology list is also provided in the nomenclature section, at the beginning, to ease the reading experience. The presented experimental results and the flexibility of the system showcase that DuraMote is a robust, durable and sustainable platform and a promising step towards real-time remote monitoring of structures, scalable, advanced SCADA systems and infrastructure networks.

2. DuraMote platform

To cover a wide range of structural health monitoring (SHM) applications, DuraMote is designed to have flexible and robust hardware platforms and modular and adaptable software features. It is also equipped with a remote firmware updating function that enables the parameters of the system to be easily adjusted for different SHM applications.

The platform consists of two types of nodes, a sensing node named Gopher, and a data aggregator named Roocas, as shown in Figures 1 and 2, respectively. Such a configuration enables the DuraMote system to build up a tiered networking system: sensing/actuation, data aggregation and integration Monitoring & Analysis (iMA) tiers as shown in Figure 3. The nodes are enclosed in a NEMA 4+ enclosure to protect them from harsh environmental conditions. In order to protect the assembled printed circuit boards against high humidity, H_2S gas and corrosion, conformal coating made of modified polyurethane resins is employed as well.

To achieve high-fidelity data regardless of SHM applications, the sensing node supports low-noise floor sensors, high-precision signal conditioning and various reliable short-range (i.e., $10-50\,\mathrm{m}$) communication methods. The Gopher node is designed with up to three

Lower Board   Upper Board

-CAN bus connection: Daisy-chained connection up to 100 nodes
-Supports low noise pigtail-type MEMS accelerometer
-Supports 4th interface for a commercial piezoelectric accelerometer
-Supports multiple sensors, such as pressure gauge, humidity sensor, temperature sensor, etc
-Mechanical dimensions (LxW): 75x50 mm

Figure 1. Design and features of Gopher.

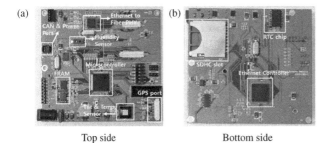

Top side   Bottom side

- Multi-communication interfaces: Wi-Fi/ Xstream/ XBee/ Eco/
  Media converter (Twisted pair to Fiber cables)
- Different topologies: P to P, P to MP, 3G network, mesh topology
- Data logging: non-volatile storage (FRAM, SDHC)
- Supports real time monitoring system: Server/Client program
- Power source: Harvesting, AC grid power, battery, supercapacitor
- Time synchronization modules: GPS, RTC, WWVB
  (atomic time broadcast)
- Mechanical dimensions (LxW): 90x100 mm

Figure 2.   Design and features of Roocas.

axes of high-precision, low-noise (5 μg/√Hz) Silicon Design's SD1221L-002 MEMS accelerometers, plus one expansion port for connection to other types of accelerometer, such as piezoelectric. One MEMS accelerometer is on the Gopher board, while the other two are of a pluggable type with sockets. In this way, the number of

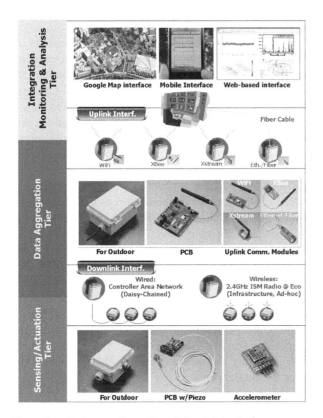

Figure 3.   System configuration of DuraMote platform.

axes can be adjusted from one to three. In addition, 8-channel analogue to digital converter (ADC) with 10-bit resolution is built in a microcontroller to connect other types of sensors, such as inclinometer, humidity, strain, gas, pressure, flow, temperature sensors, etc.

The data aggregator, Roocas, contains the processing, storage and communication links to support all data aggregation tasks. It can process data (filter, transform, compress, etc.) in software; it can log data into its flash memory card as needed and also provides several options for its data downlink and uplink (Table 1 and Figure 3).

2.1   Communication interfaces

DuraMote system supports both wired and wireless communication. Data aggregation tier bridges the communication interfaces between the downlink to the sensing/actuation tier and uplink to the iMA tier. Figure 3 displays all communication interfaces and units used in the DuraMote platform and Table 1 shows detailed specifications of both uplink and downlink interfaces (see the nomenclature section for abbreviations).

The downlink interfaces are the communication link between Roocas and Gopher, which include both a wired interface of Controller Area Network (CAN) and a wireless interface of Eco node (Chou, 2011), working as a short-range wireless communication module at a 2.4 GHz ISM band radio frequency. Since the sensor nodes in the sensing/actuation tier can establish a very dense network in a small area by connecting up to hundreds of nodes, the downlink interfaces need to support short-range communication with high-throughput to aggregate the measured data to a cluster head in data aggregation tier. For the wireless connectivity, Eco nodes can work as either a self-contained sensor node or as a wireless plug-in module, which supports short-range radio frequency communication of up to a few dozen meters at 2 Mbps. However, considering underground or underwater deployment, wireless communication is not suitable due to signal losses and high levels of attenuation. In these cases, CAN protocol can be adopted as a wired downlink communication, since it has been designed to allow communication within noisy environments and is a low-complexity, high-throughput wired bus. Furthermore, with utility power and energy harvesting interfaces, the system can distribute power to daisy-chain connected multiple sensing nodes via CAN bus, called Power over CAN, which contains two extra wires to carry power along the two for the CAN data.

The uplink interfaces of Roocas include three wireless and one wired, XStream, XBee Pro, Wi-Fi (default) and Ethernet to Fibre (i.e. 10 Base-T to 100 Base-FX media converter). The multiple uplink modules are designed as a stackable structure so that they can be easily configured to meet specific application requirements, including the

Table 1. Communication interfaces in data aggregation tier: Wireless and Wired.

Standard	XStream	XBee Pro	Wi-Fi 802.11bgn	Eco Series
			Wireless interfaces	
Frequency	902–928 MHz	902–928 MHz	2.4–2.5 GHz	2.4 GHz Shock burst
RF throughput	19.2 kbps	200 kbps	<54 Mbps	2 Mbps
Modulation type	FHSS	FHSS	OFDM with BPSK, QPSK, 16-QAM, and 64-QAM	GFSK
TX power	100 mW (20 dBm)	50 mW (+17 dBm)	15 dBm for 802.11 g/n	Programmable –18 to 0 dBm
RX sensitivity	–110 dBm (@ 9600 bps)	–100 dBm	11 Mpbs: –88.0 dBm, 54 Mpbs: –75.0 dBm	–82 dBm at 2 Mbps
Communication range	32 km @ LOS	14 km @ LOS	300 m @ LOS	10m @ LOS
Peak power consumption	210 mA @ 3.3 V	150 mA@ 5 V	260 mA @ 3.3 V	30.8 mA @ 3.3 V

Standard	Media converter		Controller area network
			Wired interfaces
Throughput & data format	LAN: 10 Base-T (10 Mbps)	Fibre: 100Base FX (100 Mbps)	1 Mbps
Communication range	LAN (~100m)	Optical (~40km)	<8 byte data frame
Remarks	Lots of commercial equipment		Designed to allow communications
	Proven technology in local area networks		within noisy environments

communication range, data rate and power management. For instance, Wi-Fi and Ethernet are apt for data streaming, while the longer-range wireless modules XBee Pro and XStream have lower data rates, than the shorter-range ones do, making them more suitable, for example, in cases where wired or wireless Internet coverage is not available. In addition, the lower data bandwidth of these longer-range interfaces qualifies them mostly for event-driven monitoring methods rather than data streaming.

2.2 High-fidelity data with flexible bandwidth

Gopher is designed with a circuit-based anti-aliasing filter, user-definable sampling rates and cut-off frequencies, and customisable digital filters as shown in Figure 4. In more detail, the anti-aliasing filter is an analogue single-order low-pass filter consisting of resistors (R), capacitors (C) and an operational amplifier. The output signal of the MEMS accelerometer is a ± 4 V differential output. This single-order electrical circuit works as the differential to single-ended conversion, as well as anti-aliasing filter with gain control. The gain of the filter can be determined by the ratio of two resistors and the cut-off frequency f_0 (3 dB frequency):

$$f_0 = \frac{1}{2\pi RC}.$$

Considering user experience, we replaced the resistors (R) of the filter circuit with digital potentiometers, enabling the cut-off frequency to be adjusted according to the frequency range of interest in each case. The advantage of the analogue single-order RC filter is that the circuitry is simpler than other high-order filters, even though the skirt of the filter is not sharp. As a result, one can easily control the cut-off frequency by digital potentiometers and avoid a possible increase of the power consumption of the sensor node due to a complex circuitry.

To compensate for the blunt skirt of the filter, a user-adjustable digital finite impulse response (FIR) filter is also available at each channel. The chipset supports a 16-bit ADC with four individually programmable 512-tap digital FIR filters (QuickFilter_technologies, 2009). The filters can be routinely designed by determining several parameters, such as pass-band ripple, pass-band upper frequency, stop-band lower frequency, etc. In addition, to ensure a flexible sensing system, the sampling rate is also adjustable. The DuraMote system can and has achieved high sampling frequencies (1.0–1.2 kHz) with simultaneous data transmission.

The connectivity for other sensors such as humidity, temperature, pressure and flow sensors is achieved by an I^{2}C (Inter-IC) interface (Philips_Semiconductors, 1995),

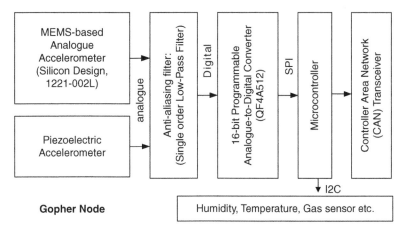

Figure 4.   Block diagram of the Gopher for signal conditioning.

or an ADC built in a microcontroller unit (MCU) of the DuraMote platform (Figure 4).

2.3   *Remote firmware updating function*

To maximise the flexibility of the system, a remote firmware updating function is an indispensable tool for bug fixes and feature upgrades. Moreover, since sensor nodes may be deployed in a large-scale area, direct programming or replacement of sensors on site can be very costly. This issue is addressed by adding a remote reprogramming function to the DuraMote platform. In this section, we describe how the remote programming approaches are designed and implemented in terms of uploading a binary image file and writing the uploaded binary image into the code execution area.

For the binary image uploading, an Extended Intel Hex format is suitable for encoding firmware images. This is because the Extended Intel Hex format allows both direct and indirect addressing and supports checksum for data consistency. Since the binary image can be uploaded either through a communication channel, such as Wi-Fi, or through an integrated file system, built upon a Secure Digital High Capacity (SDHC) card, we integrated the parser for the Extended Intel Hex format on the Gopher, which can reduce communication cost and waste of storage. In more detail, the uploaded binary file is parsed and stored in the spare area of the flash memory, integrated in the MCU. The parser converts each 32-byte ASCII data block into a 16-byte binary block and fills up the write buffer for block writes on the flash memory. During data writing on the flash memory, the integrated parser performs data size shrinking and data consistency checking.

After the binary image upload finishes, the MCU goes into reset mode. As soon as it resets, the bootstrap loader is automatically called. On the DuraMote, the bootstrap loader checks the version of the newly uploaded code image and overwrites the current application code with the new code from the image. After overwriting the application code area with the new code, the bootstrap loader controls the reset vector so that a reset event causes the program counter to start with the bootstrap loader instead of the application code when the MCU resets. When all the new code segments are successfully updated, the bootstrap loader forces a jump to the start address of the application code area so that the application can run.

3.   UCI campus experiments

In order to test the long-range wireless communication capabilities of the DuraMote platform, an experiment was set at UCI campus. The long range wireless communication network was formed with Cisco Aironet 1300 AP (Access Point), which was chosen for its ability to act as a repeater for relaying packets and for runtime configuration capabilities. We equipped it with a Cisco 4-foot-long omnidirectional high-gain antenna (12 dBi, AIR-ANT24120) and drove it with high transmission power. Two different topologies were tested, *extended star* and (linear) *multi-hop*. Figure 5 presents

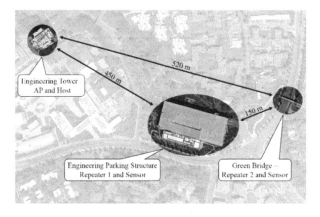

Figure 5.   Layout of the experiment.

the selected three test locations at UC Irvine campus, Engineering Tower (ET), Engineering Parking Structure (EPS) and Green Bridge (GB).

In both topologies, the host PC and the main AP were located at the top of ET. For the extended star topology case, the repeaters were at EPS and GB and were directly associated with the main AP but not with each other. For the multi-hop topology case, the EPS repeater became a relay for the GB repeater. In both cases, the DuraMote was associated with its nearest repeater, which forwarded data to the main AP at ET. For the multi-hop topology, the measured data at GB were transmitted through the EPS relay. The uplink interface between repeaters and Roocas is a Wi-Fi connection, and the network topology between them is infrastructure mode, while the downlink interface between Roocas and Gopher is daisy-chained CAN bus. Each Roocas transmitted 1000 records per second, and the AP at ET and repeaters at both EPS and GB were configured for 1–2 Mbps bandwidth and Wi-Fi Protected Access (WPA2). The host PC sent a PING request and measured the delay of PING response from the target device.

The transmission delay during the experiment is shown in Table 2 and the packet drop rate for the extended star and multi-hop topologies can be seen in Table 3. Note that the response time and packet transfer performance depend very much on the channel condition, which depends on the number of hops, communication distance and external interferences. In addition, the internal antenna used in the nodes further contributed to response time variations due to the relatively low gain. Overall, the experiment indicates DuraMote's satisfactory performance and that the multi-hop topology can effectively extend the communication distance with low packet drop rates.

Table 2. Transmission delay data.

Device	Min (ms)	Max (ms)	Average (ms)
Repeater 1	2	57	9
Repeater 2 (AP associated)	2	35	8
Repeater 2 (relayed)	4	25	9
Sensor at Parking Structure	14	67	48
Sensor at Green Bridge	7	96	21

Table 3. Packet transfer performance.

Topologies		Sensor at Parking Structure to host PC	Sensor at Green Bridge to host PC
Extended Star	#Tx Packet	135,405	103,738
	#Rx Packet	135,153	100,001
	Drop rate	0.2%	3.6%
Multi-Hop	#Tx Packet	388,526	355,808
	#Rx Packet	386,123	352,515
	Drop rate	0.6%	0.9%

3.1 Established network at UCI

To test the long-term performance of DuraMote, its reliability and robustness, a small network is created at UCI. As seen in Figure 6, five nodes are installed at five different buildings and data are gathered at a server location, located at AIRB building. The sampling rate is 100 Hz and the Roocas directly access the Internet, taking advantage of the available wireless network in UCI. Building number 3 is Calit2 building and DuraMote is placed at its roof and in some proximity with a wired Kinemetrics accelerometer installed by USGS at the ceiling of the fourth floor of the building. Since its installation, in December 2011, the network works uninterruptedly and continuously monitors the buildings. Several earthquakes have occurred during this time and have been successfully recorded by the system. Table 4 indicates some earthquake characteristics of the most significant recorded ones.

One of the earthquakes, the closest to campus at Laguna Niguel (April 2012), is also recorded by the USGS wired Kinemetrics sensor. A comparison of the two signals can be seen in both frequency and time domain in Figures 7 and 8, respectively. Although the sensors are at different locations and their sampling frequency and technology are different, results are in close agreement validating DuraMote's accuracy and performance in situ.

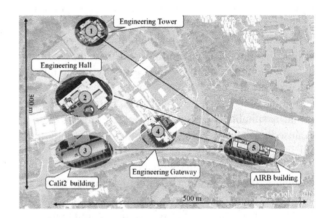

Figure 6. Campus network.

Table 4. Recorded earthquakes.

Date	Earthquake	Mw	Distance from UCI (km)
23 April 2012	Laguna Niguel	3.9	19
12 June 2012	1st Yorba Linda	4.1	30
7 August 2012	2nd Yorba Linda	4.5	30
8 August 2012	3rd Yorba Linda	4.5	30
26 August 2012	1st Brawley	5.4	225
26 August 2012	2nd Brawley	5.5	225
29 August 2012	4th Yorba Linda	4.1	30
15 May 2013	S. of Rancho Palos Verdes	4.0	55

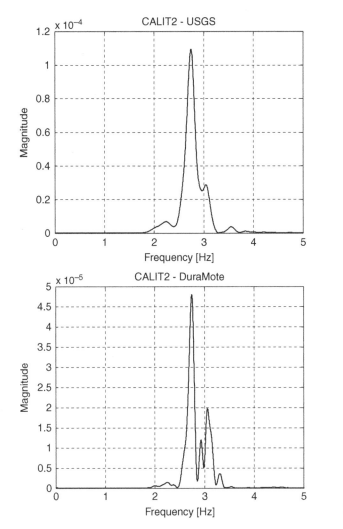

Figure 7.   Frequency domain, Laguna Niguel earthquake.

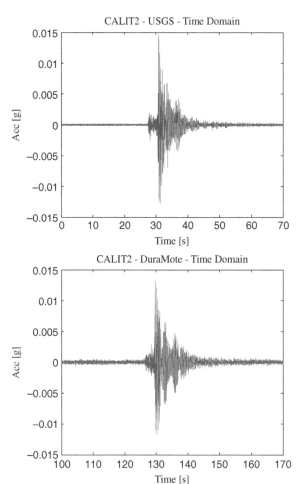

Figure 8.   Time domain, Laguna Niguel earthquake.

The frequency domain results are also consistent with system identification results for this building in Ulusoy, Feng, and Fanning (2011). An important note at this point is also the fact that for all the other earthquake events in Table 4, the USGS sensor did not record the signals, losing valuable information from the records, because it does not monitor continuously and the vibration intensity was not large enough to trigger its recording mechanism.

4.   Long-span bridges applications

4.1   Vincent Thomas bridge

In order to test the performance of the wireless network in a noisy environment, such as a bridge with steel girders, the Vincent Thomas bridge was used as a test bed. The Vincent Thomas bridge is a four-lane suspension bridge located in San Pedro, California, spanning over the main channel of Los Angeles Harbor and it is a part of Seaside Freeway and State Route 47. The bridge was designed by the bridge department of the California Department of Transportation in 1960 and opened to traffic in 1963. Currently, it is the 19th longest suspension bridge in the USA and the third in California. The bridge's deck is supported by two towers resting on steel piles. The main span is 457 m long and the side spans are 154 m each.

Three DuraMote platforms were used in this test and placed in a symmetrical configuration, as seen in Figure 9. Each DuraMote consisted of one Roocas and two Gophers. We set WDS-enabled (Wireless Distribution System) APs (Buffalo WHR-HP-G300N 802.11 n routers) approximately 100 m apart, capable of 150 Mbps throughput, to form a local wireless-relaying infrastructure using relatively small and light antennas. Totally, three Roocas, six Gophers and five WDS APs were deployed on the bridge for 2 h, as seen in Figure 9, and the sampling frequency during this experiment was 1 kHz. Table 5 shows the number of packets transmitted by each Roocas on the bridge and the number of packets received by one single AP during the experiment. The packet drop rate can be also seen in Table 5 and is below 0.5% for this network topology. This is considered reliable enough for high-fidelity data acquisition.

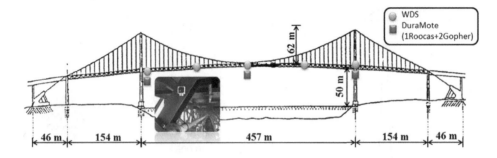

Figure 9. Deployment of DuraMote on the Vincent Thomas Bridge and Gopher nodes at one location.

Table 5. Packet drop rate.

	Transmitted samples	Received samples	Drop rate (%)
Sensor 1	649,580	648,315	0.19
Sensor 2	639,192	636,272	0.45
Sensor 3	626,261	624,047	0.35

Based on the recorded data from the three sensors, frequency domain results have been computed based on the Frequency Domain Decomposition method, (Brincker, Zhang, & Andersen, 2001), and the four basic eigenfrequencies (relating to vertical modes), as seen in Figure 10, were identified successfully in agreement with system identification results in (Ulusoy, 2011). More specifically, the estimated eigenfrequencies in Figure 10 are reported as 0.228, 0.371, 0.478 and 0.818 Hz. The respective estimations in (Ulusoy, 2011) are reported as 0.232, 0.364, 0.461 and 0.807 Hz and are based on subspace system identification algorithms and two earthquake induced vibration datasets (2008 Chino Hills and 2009 Inglewood earthquakes), recorded by the sensor network on the bridge that is maintained by the California Geological Survey. Further

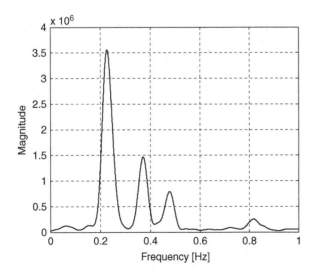

Figure 10. Vincent Thomas bridge eigenfrequencies.

details about this experiment and respective results can be found in Kim, Torbol, and Chou (2013).

4.2 Hwamyung bridge installation

The DuraMote system has been tested and permanently installed on the Hwamyung bridge, which is a cable stayed bridge between the city of Busan and the town of Gimhae, in South Korea. The description in this paper relates to a temporary DuraMote installation on the bridge. For the permanent sensor network details, with a sampling frequency of 100 Hz, the interested reader is referred to Torbol, Kim, Chien, and Shinozuka (2013). The Hwamyung bridge is a newly built cable stayed bridge with a main span of 270 m. The top of the towers are 65 m above the deck. The bridge was instrumented with 10 DuraMote nodes and the Gopher-Roocas connection can be seen in Figure 11. Seven Gophers were placed on the deck, with two accelerometers each to measure the vertical and transversal deck vibrations. Two Gophers were placed on top of the towers, with two accelerometers each to measure the longitudinal and transversal tower vibrations. Finally, four Gophers were placed on the cables.

The experiment lasted several months and data were transmitted continuously through Wi-Fi to the base station, which was placed inside the tower on the Gimhae side. The sampling frequency was set as high as 450 Hz in order to produce heavy packet traffic and test the communication performance. The base station had Internet access through a 3G/4G connection and it could be accessed remotely. During the time of the experiment, the bridge was subjected to heavy rain, wind and a typhoon, but the system continued to aggregate acceleration data proving its robustness. Overall, several system capabilities, such as high data-fidelity, system flexibility, high wireless throughput, local data saving, etc., were successfully tested during this experiment. The natural frequencies and mode shapes of the bridge were also effectively identified and are in good agreement with analytical results, obtained from a Finite Element model of the bridge, as seen in Table 6. Further details about analytical and experimental results and sensor deployment can be seen in Torbol, Kim, and Shinozuka (2013).

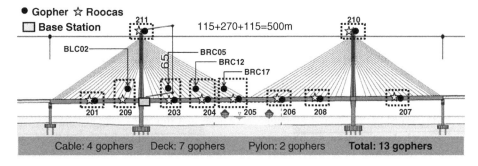

Figure 11. Hwamyung bridge geometry and sensor layout.

Table 6. Identified natural frequencies of the Hwamyung bridge.

Experimental ω_n (Hz)	Analytical ω_n (Hz)	Direction
0.413	0.476	Transversal
0.465	0.484	Vertical
0.752	0.833	Vertical
0.797	0.797	Transversal
1.072	*	Vertical
1.134	1.165	Vertical
1.276	1.219	Vertical
1.478	1.238	Transversal
1.522	1.561	Vertical
2.270	2.299	Vertical
2.475	2.112	Torsional

*Not present in the analytical model.

5. Pipeline systems monitoring

As mentioned in the introduction, apart from single structures the DuraMote platform is designed and tested in order to monitor continuously and in real-time, spatially distributed, large-scale civil infrastructure systems, with emphasis on pipeline networks and SCADA systems. As already explained, traditional SCADA systems have sensors only in key locations of a pipeline network, such as main and pump stations. Yet, they can still in principle be used for damage detection purposes as well (Zhang, 1996). However, the usual sparseness of the data-acquisition locations within the network makes it difficult to detect the damage with accuracy and rapidity in practice. DuraMote has the ability to monitor remotely, continuously and in real-time large spatial areas, can handle multiple, different sensor units and requires low power demand for sensing and transmission. Thus, wireless monitoring of extensive areas of a network by placing sensors along the pipelines is becoming a feasible task. The advanced attributes of such a system provide new ways of identifying damage events and unusual, unwanted network behaviour that were impossible without this new technology.

A variety of sophisticated methods have been suggested in the literature for leak and pipe break detection (Colombo, Lee, & Karney, 2009). The majority of them are based on pressure and/or velocity flow monitoring (Liggett & Chen,

1994; Misiunas, Vitkovsky, Olsson, Simpson, & Lambert, 2005), and acoustic emission techniques, (Hunaidi & Chu, 1999). Procedures based on vibration data are also present (Evans, Blotter, & Stephens, 2004; Gao, Brennan, Joseph, Muggleton, & Hunaidi, 2005; Hunaidi & Chu, 1999), relying only on acceleration recordings during steady-state flow. Although all these techniques, as well as traditional SCADA approaches, would benefit from a densely instrumented network, the focus in this work is only on a newly suggested, simple, practical and non-invasive method, for the often encountered problem of pipe bursts. This technique is been lately investigated by Shinozuka and co-workers and can fully exploit the advanced capabilities of DuraMote, for identification of pipe breaks.

The suggested damage identification method is based on pipe vibrations and acceleration data recorded during transient flows, caused by abrupt pipe bursts or sudden malfunctions. Such incidents generate pressure waves, propagating along the pipeline, causing significant hydraulic transients (Misiunas et al., 2005; Srirangarajan et al., 2013). The dynamic variation of flow pressure induces significant pipe vibrations. Hence, acceleration recordings in time, by the MEMS accelerometers, installed at appropriate locations, can indicate the existence and location of damage by simply observing the magnitude of the signals in time domain and their cross-correlation time plots (Papakonstantinou et al., 2011; Shinozuka, Chou, Kim, Kim, Karmakar, et al., 2010).

In the rest of this section, the relationship between pressure variations and pipe acceleration is analysed and explained. Its exploitation for damage identification of a pipe water network becomes apparent as well. Simulation results from a virtual utility network are firstly, briefly discussed. Results indicate that pressure variations in time (water head gradient) could identify the vicinity of a pipe rupture. Experimental results are then shown, to illustrate the flow pressure acceleration connection. A scaled pipe network was mainly used for this purpose, where pressure gauges and accelerometers were installed and valuable data were recorded during an induced transient flow. The interaction between pressure variations and pipe vibrations is also theoretically explained in Papakonstantinou et al. (2011).

In Shinozuka and Dong (2005), the computer code HAMMER was used to generate transient water head time histories. The water delivery network system used is shown in Figure 12. The system consists of 2 reservoirs, 1 pump, 1 valve, 38 nodes and 54 pipe links covering an area approximately $40\,km \times 60\,km$. The initial hydraulic boundary conditions are such that reservoirs supply water to the network which is consumed, in steady state, at consumption nodes. In this virtual but realistic network, a damage scenario is considered, where a pipe break is assumed to occur at the mid-point of link P111 between joints J9 and J11 (Figure 12). The water head gradient, D, is defined as $D = |\dot{H}|$ where H water head and dot represent differentiation with respect to time. The absolute maximum value of D computed over the first 30 seconds after the pipe break is used as D value at each joint, and its spatial distribution over the entire network is plotted for this scenario in Figure 12. Figure 12 shows significant hydraulic transients and water head gradient, after rupture, at nodes closely located to damage. Thus, pressure

recordings at different network nodes which firstly demonstrate, in time, a high value of D can easily reveal the closer to the rupture nodes. D values in Figure 12 are artificial except at the location of joints where they have been calculated. This contour map, however, facilitates effective automated identification of the links with locally maximum D values at their ends and can be achieved in near real-time. By using a cross-correlation analysis in time, of different node recordings, and utilising wave propagation speeds, damage location may also be more accurately identified.

Following the same concept and since flow pressure gradient and pipe acceleration are strongly correlated quantities, as is shown soon after, accelerometer recordings, along the pipeline, can be utilised for damage identification purposes. The major advantage of utilising acceleration data is that Gophers with low-cost MEMS accelerometers can be simply and non-invasively installed on the external pipe surfaces, enabling large-scale deployment at an affordable cost, even in spatially extensive water pipe networks. Obviously, a combination of flow sensors (pressure gauges, flow meters, etc.) and accelerometers will be able to identify

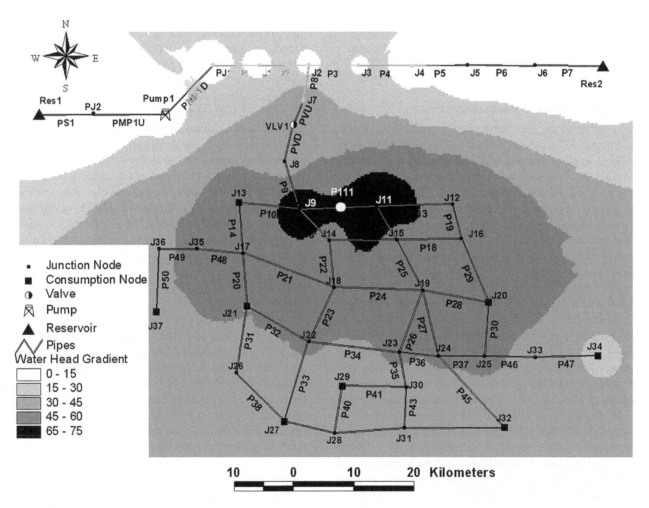

Figure 12. Contour map of water head gradient.

damage events more reliably and accurately, since several detection methods can be then utilised and the fusion of sensors reduces false alarm potentials. In case such configurations are chosen, the DuraMote platform can also support them, as already explained.

5.1 Simple laboratory experiment

A small laboratory test was conducted to check the correlation between water pressure and pipe acceleration. A single, horizontal water pipe with two valves was set up. The release valve at the middle of the pipe (Figure 13) was introduced in order to simulate a rupture point, while the outlet valve controlled the pressure in the pipe by adjusting its opening. The inlet was connected to the utility water supply line. Pipe material was PVC 40, with an inner diameter of 1" (0.0254 m) and wall thickness of 0.25" (0.00635 m). The experimental setup included six pressure sensors, five unidirectional piezoelectric accelerometers and four three-dimensional MEMS sensors on the Gophers, as seen in Figure 13. The "A's" in this figure indicate accelerometer locations (same for Gophers and piezoelectric accelerometers) and their numbering and arrow directions relate to the piezoelectric ones. The sampling rate for all sensors was set equal to 1250 Hz.

During the experiment, different pressure change conditions were observed for different induced boundary conditions, e.g. a valve opening or use of water inside the building. As can be also seen in Papakonstantinou et al. (2011), the pressure change was always coupled with an acceleration spike, indicating significant pipe vibrations. An example of the obtained results is shown in Figure 14. For this example, the release valve was suddenly opened twice to simulate the rupture events, at around 22 s and 94 s. An interesting observation is also that during valve closing, around 46 s and 124 s, the acceleration magnitude is not as large as during the valve-opening case, although the pressure difference is nearly identical in both cases. The reason is that the valve closing speed was approximately 10 times slower than the opening one, affecting the pressure gradient.

5.2 Scaled water network

In order to further investigate experimentally the pressure wave propagation and pipe acceleration for damage

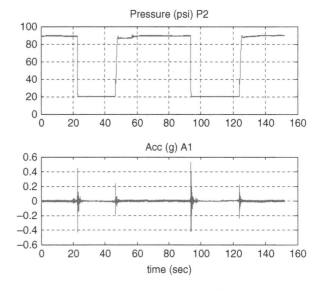

Figure 14. Pressure and acceleration time histories.

detection purposes, a scaled water network has been developed, near UCI campus, at Rattlesnake Reservoir location as shown in Figure 15. The system consists of 4" (0.1016 m) diameter PVC pipes. The core of the system is around 20 m by 20 m with valves placed throughout the system to open and close water pathways and change its topology. The vertical pipes seen in Figure 15 were placed in case the pipe network is covered with soil in the future. Water is supplied to the network either directly from a street line or by a pump system installed together with a 2,000 gallon tank. To simulate a rupture event, as in Figure 16, rubber plates were used. As presented in the photographs, 3/8" thick rubber plates can be held in place by a ring and can act as an impermeable membrane. The ring can be tightened such that at a certain pressure, when the friction of the rubber plates is overcome by the internal pressure, the plates eject.

In Figure 17, a sketch of the experimental facility can be seen together with accelerometers and pressure gauges locations, for one of the tests performed. Fourteen piezoelectric accelerometers and 14 unidirectional

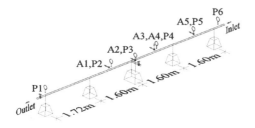

Figure 13. Experimental setup and sensor layout.

Figure 15. Scaled water network.

Figure 16. An indicative rupture event and the rupture mechanism with a ring (grey) and rubber plates.

MEMS sensors were used (at the same locations), together with 10 pressure sensors. In total, 14 Gopher nodes and seven Roocas were used. The sampling rate for all sensors was set equal to 1000 Hz. In order to aggregate all data in real-time fashion, high-throughput wireless technology was required. For this experiment, Roocas equipped with Wi-Fi 802.11 n was employed to transmit the measured data to our server system. For the network topology between data aggregators and sensing nodes, one data aggregator was daisy-chained to two sensing nodes via CAN bus. Due to the multiple and high throughput wireless channels, data I/O began to bottleneck and slow the system's performance. This issue was addressed by the close file action for each wireless channel in a Time

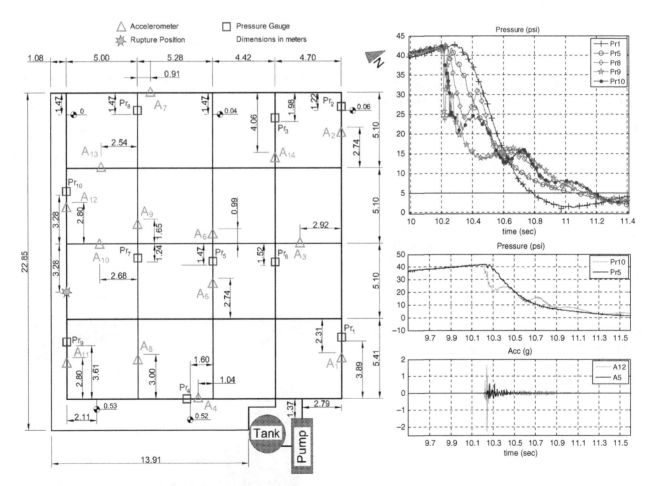

Figure 17. Sketch of the experimental water network, pressure time histories after a rupture event and pressure and pipe acceleration time histories comparison between two pairs of sensors.

Division Multiple Access (TDMA) style, which relieved the heavy load of closing the file descriptor and syncing all data back into the disk buffer.

In Figure 17, pressure time histories after a rupture event and pressure and pipe acceleration time histories comparison between two pairs of sensors can be seen. Locations closer to the damage exhibit more intense transient effects that are accordingly reflected in the pressure time histories. The dynamic difference in pressure variations can be easily identified through pipe accelerations as seen in the comparison part of Figure 17. Time delays due to wave propagations can be noticed as well. Figure 18 illustrates the clear connection between the second derivative of the pressure in time and the pipe acceleration from four different pairs of sensors and multiple rupture events. The rupture location and the channel numbering for both pressure gauges and accelerometers are consistent with Figure 17. Finally, Figure 19 shows acceleration and normalised pressure gradient contour maps, in accordance with Figure 12. The black dots in this figure represent the sensor locations for this experiment and the star represents the rupture position. Findings based on Figure 12, concerning the rupture location, are valid in Figure 19 as well. In addition, the acceleration contour is also shown and can indeed also be used for damage identification purposes.

In general, experimental results clearly proved the relationship between pressure transients and pipe acceleration. Both types of measurement can therefore be used for rupture localisation, either independently or combined. Further instrumentation of a spatially large, real network will provide useful insight into the advantages and disadvantages of each approach and is a direction worth

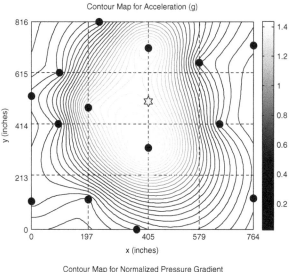

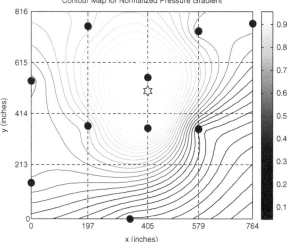

Figure 19. Pipe acceleration and normalised pressure gradient contour maps.

pursuing. Several efforts have already been made by the authors on realfield applications. An example, for instance, is a site administered by Orange County Sanitation District (OCSD), where solar-powered DuraMotes are installed on pipes, together with a 20 W multi-crystalline silicon solar panel and EscaCap, the supercapacitor-based harvesting circuitry with built-in MPPT. The system has a sampling frequency of 1 kHz and a cellular modem connection (Kim et al., 2013).

6. Conclusions

The concept of real-time, remote monitoring of structures and spatially distributed, large-scale infrastructure systems is presented in this work. The DuraMote platform, with the abilities, among others, to cover large areas and to handle different sensor units has been described as well. Field experiments and real-life applications with buildings, bridges and pipeline networks validate the performance of

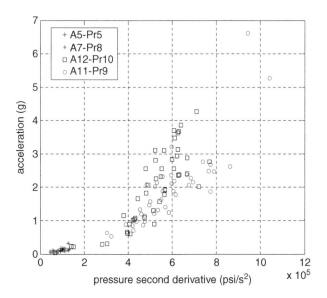

Figure 18. Pressure second derivative and pipe acceleration connection from four different pairs of sensors and multiple rupture events.

DuraMote and the feasibility for expansion of the existing concept of structural health monitoring, into remote, real-time, continuous and permanent performance monitoring of spatially extended systems. Particularly, pipeline networks could benefit a lot from this technology, due to new and enhanced damage identification methods, as is explained in detail. Overall, the current work sets firm foundations for the implementation of advanced, flexible, scalable and robust new generation Supervisory Control and Data Acquisition (SCADA) systems for structural networks.

Acknowledgements

The authors would like to thank Dr Sergio Carnalla for the support during the experiments, as well as the research group from the Department of Electrical Engineering and Computer Science at UCI including Prof. Pai H. Chou, Mr Ting-Chou Chien and Dr Eun-bae Yoon for their contributions concerning the development of DuraMote.

Funding

This study was carried out under National Institute of Standards and Technology (NIST) grant. Their support is deeply appreciated. The partners on this project include Earth Mechanic Inc. (EMI), Irvine Ranch Water District (IRWD), Orange County Sanitation District (OCSD), Santa Ana Water Project Authority (SAWPA) and Pacific Advanced Civil Engineering (PACE). The authors thank all these partners for their support and contributions. The authors also acknowledge the support provided by National Science Foundation for the early conceptual development of this study. The support of Hyundai for the experiment on the Hwamyung bridge is greatly appreciated.

Notes

1. Email: shino@uci.edu
2. Email: kpapakon@uci.edu
3. Email: mtorbol@unist.ac.kr

References

Allen, M., Preis, A., Iqbal, M., Srirangarajan, S., Lim, H.B., Girod, L., & Whittle, A.J. (2011). Real-time in-network distribution system monitoring to improve operational efficiency. *Journal American Water Works Association (AWWA), 103,* 63–75.

Brincker, R., Zhang, L.M., & Andersen, P. (2001). Modal identification of output-only systems using frequency domain decomposition. *Smart Materials and Structures, 10,* 441–445.

Chou, P.H. (2011). Eco node. Retrieved from: http://newport.eecs.uci.edu/~chou/research.html [Accessed March 2014].

Colombo, A.F., Lee, P., & Karney, B.W. (2009). A selective literature review of transient-based leak detection methods. *Journal of Hydro-environment Research, 2,* 212–227.

EPA (2010). *Addressing the challenge through innovation.* Environmental Protection Agency, US. Retrieved from http://

nepis.epa.gov/Adobe/PDF/P100EQ6Q.pdf [Accessed March 2014].

Evans, R., Blotter, J., & Stephens, A. (2004). Flow rate measurements using flow-induced pipe vibration. *ASME Journal of Fluids Engineering, 126,* 280–285.

Gao, Y., Brennan, M., Joseph, P., Muggleton, J., & Hunaidi, O. (2005). On the selection of acoustic/vibration sensors for leak detection in plastic water pipes. *Journal of Sound and Vibration, 283,* 927–941.

Ho, D.-D., Lee, P.-Y., Nguyen, K.-D., Hong, D.-S., Lee, S.-Y., Kim, J.-T., Shin, S.-W., Yun, C.-B., & Shinozuka, M. (2012). Solar-powered multi-scale sensor node on Imote2 platform for hybrid SHM in cable-stayed bridge. *Smart Structures and Systems, 9,* 145–164.

Hunaidi, O., & Chu, W. (1999). Acoustical characteristics of leak signals in plastic water distribution pipes. *Journal of Applied Acoustics, 58,* 235–254.

Jin, Y., & Eydgahi, A. (2008). Monitoring of distributed pipeline systems by wireless sensor networks. In *Proceedings of the 2008 IAJC-IJME International Conference,* November 17–19. Nashville, TN: IAJC.

Kim, S., & Chou, P.H. (2011). Energy harvesting by sweeping voltage-escalated charging of a reconfigurable supercapacitor array. In *International Symposium on Low Power Electronics and Design (ISLPED),* August 1–3. Fukuoka, Japan: ACM press.

Kim, S., Chou, P., & Shinozuka, M. (2011). Smart wireless sensor system for lifeline health monitoring under a disaster event. In *Proceedings of SPIE Vol. 7983-77, Nondestructive Characterization for Composite Materials, Aerospace Engineering, Civil Infrastructure, and Homeland Security.* San Diego, CA: SPIE.

Kim, S., No, K.-S., & Chou, P.H. (2011). Design and performance analysis of supercapacitor charging circuits for wireless sensor nodes. *IEEE Journal on Emerging and Selected Topics in Circuits and Systems, 1,* 391–402.

Kim, Y., Schmid, T., Charbiwala, Z.M., Friedman, J., & Srivastava, M.B. (2008). NAWMS: Nonintrusive autonomous water monitoring system. In *The 6th ACM Conference on embedded networked sensor systems (SenSys 2008),* November 04–07. Raleigh, NC: ACM press.

Kim, S., Torbol, M., & Chou, P. (2013). Remote structural health monitoring systems for next generation SCADA. *Smart Structures and Systems, 11,* 511–531.

Ko, J.M., & Ni, Y.Q. (2005). Technology developments in structural health monitoring of large-scale bridges. *Engineering Structures, 27,* 1715–1725.

Lewis, F.L. (2004). Wireless sensor networks. In D.J. Cook & S. K. Das (Eds.), *Smart environments: Technologies, protocols, and applications.* New York: John Wiley.

Liggett, J.A., & Chen, L.C. (1994). Inverse transient analysis in pipe networks. *ASCE Journal of Hydraulic Engineering, 120,* 934–955.

Lynch, J.P., Partridge, A., Law, K.H., Kenny, T.W., Kiremidjian, A.S., & Carryer, E. (2003). Design of piezoresistive MEMS-based accelerometer for integration with wireless sensing unit for structural monitoring. *ASCE Journal of Aerospace Engineering, 16,* 108–114.

Lynch, J.P., & Loh, K.J. (2006). A summary review of wireless sensors and sensor networks for structural health monitoring. *The Shock and Vibration Digest, 38,* 91–130.

Misiunas, D., Vitkovsky, J., Olsson, G., Simpson, A., & Lambert, M. (2005). Pipeline break detection using pressure transient monitoring. *ASCE Journal of Water Resources Planning and Management, 131,* 316–325.

Pakzad, S.N., Fenves, G.L., Kim, S., & Culler, D.E. (2008). Design and implementation of scalable wireless sensor network for structural monitoring. *ASCE Journal of Infrastructure Systems, 14*, 89–101.

Papakonstantinou, K.G., Shinozuka, M., & Beikae, M. (2011). Experimental and analytical study of water pipe's rupture for damage identification purposes. In *Proceedings of SPIE Vol. 7983, Nondestructive Characterization for Composite Materials, Aerospace Engineering, Civil Infrastructure, and Homeland Security*. San Diego, CA: SPIE.

Park, G., Rosing, T., Todd, M.D., Farrar, C.R., & Hodgkiss, W. (2008). Energy harvesting for structural health monitoring sensor networks. *ASCE Journal of Infrastructure Systems, 14*, 64–79.

Philips_Semiconductors (1995). The I2C bus and how to use it. Retrieved from: http://www.i2c-bus.org/fileadmin/ftp/i2c_bus_specification_1995.pdf [Accessed March 2014].

QuickFilter_technologies (2009). QF4A512 datasheet. Retrieved from: http://www.quickfiltertech.com/files/QF4A512revD8.pdf [Accessed March 2014].

Rice, J.A., Mechitov, K., Sim, S.-H., Nagayama, T., Jang, S., Kim, R., Spencer, B.F., Agha, G., & Fujino, Y. (2010). Flexible smart sensor framework for autonomous structural health monitoring. *Journal of Smart Structures and Systems, 6*, 423–438.

Rice, J.A., Mechitov, K.A., Sim, S.H., Spencer, B.F., & Agha, G.A. (2011). Enabling framework for structural health monitoring using smart sensors. *Structural Control and Health Monitoring, 18*, 574–587.

Shinozuka, M., Chou, P., Kim, S., Kim, H., Yoon, E., Mustafa, H., Karmakar, D., & Pul, S. (2010). Nondestructive monitoring of a pipe network using a MEMS-based wireless network. In *Proceedings SPIE Vol. 7649, Nondestructive Characterization for Composite Materials, Aerospace Engineering, Civil Infrastructure, and Homeland Security*. San Diego, CA: SPIE.

Shinozuka, M., Chou, P., Kim, S., Kim, H., Karmakar, D., & Lu, F. (2010). Non-invasive acceleration-based methodology for damage detection and assessment of water distribution system. *Smart Structures and Systems, 6*, 545–559.

Shinozuka, M., & Dong, X. (2005). Evaluation of hydraulic transients and damage detection in water system under a disaster event. In *Proceedings of the 3rd US-Japan Workshop on Water System Seismic Practices*. Kobe: AWWA Research Foundation.

Shinozuka, M., Park, C., Chou, P.H., & Fukuda, Y. (2006). Real-time damage localization by means of MEMS sensors and use of wireless data transmission. In *Proceedings of SPIE Vol. 6178, Nonintrusive Inspection, Structures Monitoring, and Smart Systems for Homeland Security*. San Diego, CA: SPIE.

Song, L. (2012). Few oil pipeline spills detected by much-touted technology. Retrieved from: http://insideclimatenews.org/print/16594 [Accessed March 2014].

Srirangarajan, S., Allen, M., Preis, A., Iqbal, M., Lim, H.B., & Whittle, A.J. (2013). Wavelet-based burst event detection and localization in water distribution systems. *Journal of Signal Processing Systems, 72*(1), 1–16.

Stoianov, I., Nachman, L., Madden, S., & Tokmouline, T. (2007). PIPENET: A wireless sensor network for pipeline monitoring. In *Proceedings of the International Symposium on Information Processing in Sensor Networks (IPSN '07)*, April 25–27. Cambridge, MA: IEEE.

Torbol, M., Kim, S., Chien, T.-C., & Shinozuka, M. (2013). Hybrid networking sensing system for structural health monitoring of a concrete cable-stayed bridge. In *Proceedings of SPIE Vol. 8692 86920C, Sensors and Smart Structures Technologies for Civil, Mechanical, and Aerospace Systems*. San Diego, CA: SPIE.

Torbol, M., Kim, S., & Shinozuka, M. (2013). Long term monitoring of a cable stayed bridge using DuraMote. *Smart Structures and Systems, 11*, 453–476.

Ulusoy, H.S. (2011). *Applications of system identification in structural dynamics: A realization approach*. PhD thesis. University of California Irvine, CA.

Ulusoy, H.S., Feng, M.Q., & Fanning, P.J. (2011). System identification of a building from multiple seismic records. *Earthquake Engineering & Structural Dynamics, 40*, 661–674.

Xu, N., Rangwala, S., Chintalapudi, K.K., Ganesan, D., Broad, A., Govindan, R., & Estrin, D. (2004). A wireless sensor network for structural monitoring. In *Proceedings of the 2nd international conference on embedded networked sensor systems (SenSys '04)*. New York, NY: ACM.

Yoon, S., Ye, W., Heidemann, J., Littlefield, B., & Shahabi, C. (2011). SWATS: Wireless sensor networks for steamflood and waterflood pipeline monitoring. *IEEE Journal of Network, 25*, 50–56.

Zhang, J. (1996). Designing a cost effective and reliable pipeline leak detection system. In *Pipeline Reliability Conference*, November 19–22. Houston, TX: Gulf Publishing Company.

Index

INDEX